Data Mining for Co-Location Patterns

Data Mining for Co-Location Patterns

Principles and Applications

Guoqing Zhou

CRC Press is an imprint of the
Taylor & Francis Group, an **informa** business

First edition published 2022
by CRC Press
6000 Broken Sound Parkway NW, Suite 300, Boca Raton, FL 33487–2742

and by CRC Press
2 Park Square, Milton Park, Abingdon, Oxon, OX14 4RN

CRC Press is an imprint of Taylor & Francis Group, LLC

Library of Congress Cataloging-in-Publication Data
Names: Zhou, Guoqing, 1965– author.
Title: Data mining for co-location patterns : principles and applications / Guoqing Zhou.
Description: First edition. | Boca Raton, FL : CRC Press, 2022. | Based on author's second Ph.D. dissertation completed in Virginia Tech, Blacksburg, Virginia, USA, 2001. | Includes bibliographical references and index.
Identifiers: LCCN 2021039989 | ISBN 9780367654269 (hbk) | ISBN 9780367688660 (pbk) | ISBN 9781003139416 (ebk)
Subjects: LCSH: Cluster analysis—Data processing. | Pattern perception—Data processing. | Data mining. | Geospatial data—Computer processing. | Business intelligence—Data processing.
Classification: LCC QA278.55 .Z46 2022 | DDC 006.3—dc23/eng/20211027
LC record available at https://lccn.loc.gov/2021039989

ISBN: 978-0-367-65426-9 (hbk)
ISBN: 978-0-367-68866-0 (pbk)
ISBN: 978-1-003-13941-6 (ebk)

DOI: 10.1201/9781003139416

Typeset in Times
by Apex CoVantage, LLC

Contents

Preface

The proliferation, ubiquity, and increasing power of various types of sensors, wire(less) communication, and advanced information technology have dramatically increased the capabilities of data collection, storage, and manipulation ability, with the result that various types of data sets have dramatically grown in volume and complexity.

Data mining is the process of finding anomalies, patterns, correlations, and new insights hidden inside large data sets to predict future trends. Data mining has been widely applied in many fields, for example, retailers, banks, manufacturers, telecommunications providers, and insurers, among others, who are using data mining to discover relationships among everything from price optimization, promotions, and demographics to how the economy, risk, competition, and social media are affecting their business models, revenues, operations, and customer relationships. It has been demonstrated that the more complex the data sets collected, the more potential there is to uncover relevant insights.

The objective of this book is to advance readers' data mining techniques through focusing on a special topic, co-location pattern mining. This book gives a comprehensive description of the principles of co-location pattern mining and its applications in geoinformatics, such as spatial analysis, spatial decision making, and remotely sensed image classification. The content of this book has been specially deliberated upon, from fundamentals of co-location definitions, to theory of co-location, to co-location decision tree, to maximum variance unfolding (MVU) co-location, to maximal instance algorithm for co-location, and to negative co-location and can therefore be used as a textbook of senior undergraduate, and graduate courses and as a reference for instructors, researchers, scientists, engineers, and professionals in academia and governmental and industrial sectors. With investigations of the published books relevant to data mining, none of them has only concentrated on the special topic of co-location data mining. Another objective of this book is therefore to fill the gap.

Several of the apparent features of the book that I believe will enhance its value as a textbook and as a reference book are as follows: First, this book only concentrates on a special topic, co-location mining pattern; that is, this textbook goes beyond the traditional focus on a wide spectrum of data mining. Second, this book emphasizes the principles, methods, and wide applications of co-location pattern mining, of which each topic is clearly explained and illustrated by detailed examples. Third, co-location pattern mining is an emerging field and one of the cutting-edge topics of data science and is able to automatically dig co-location patterns from all types of data, with applications ranging from scientific discovery to business intelligence and analytics. Fourth, the content of this book specially deliberates upon co-location definitions (fundamentals), the theory of co-location (basic theory), the co-location decision tree (middle advanced theory), the MVU co-location decision tree (junior advanced theory), and maximum instance co-location pattern mining (senior advanced theory). Such a deliberated chapter arrangement is to help students completely understand the principle of co-location pattern mining step by step.

The book significantly reflects new developments in and case applications of data mining in order to meet societal and economic challenges. I believe that readers will be very impressed and interested in co-location pattern mining. Co-location pattern mining reflects and represents the emergence of the most current data mining theory and technologies.

It is my hope that this book can be read in varying depths by readers with data science backgrounds.

Guoqing Zhou
July 2021

Acknowledgments

This book is based on my second PhD dissertation completed at Virginia Tech, Blacksburg, Virginia, USA. A special thanks to my supervisor, Professor Linbing Wang, for his impetus to write about and advise on this subject. Also many thanks to my graduate students, Qi Li and Zhenyu Wang, for their work on the maximum instance co-location pattern mining and negative co-location pattern mining, and Rongting Zhang on maximum variance unfolding (MVU)-co-location pattern mining. In addition, I would like to specially thank my research assistant, Ms. Xia Wang, who has been involved in various aspects of this book, such as format editions.

Finally, I should very much like to thank my wife, Michelle M. Xie, and two sons, Deqi Zhou and Mitchell Zhou. I thank them for those occasional times that I did dedicate to work on the manuscript rather than spend more time with them.

Guoqing Zhou
July 2021

Author Biography

Guoqing Zhou received his first PhD from Wuhan University, Wuhan, China, in 1994 and his second PhD from Virginia Tech at Blacksburg, Virginia, USA, in 2001. He was a visiting scholar at the Department of Computer Science and Technology, Tsinghua University, Beijing, China, and a postdoctoral researcher at the Institute of Information Science, Beijing Jiaotong University, Beijing, China, from 1994–1996. He continued his research as an Alexander von Humboldt Fellow at the Technical University of Berlin, Berlin, Germany, from 1997–1998 and afterward became a postdoctoral researcher at the Ohio State University, Columbus, OH, USA, from 1998 to 2000. Later he worked at Old Dominion University, Norfolk, VA, USA, as an assistant professor, associate professor, and full professor in 2000, 2005, and 2010, respectively. He is currently professor at the Guilin University of Technology, Guilin, China. He is author of five books and has published more than 300 papers in peer-reviewed journals and conference proceedings.

1 Introduction

1.1 BACKGROUND

With the rapid development of science and technology, a large amount of data has been produced in all daily lives and all fields. These data are not only various, but also stored with different types of formats. However, a large number of useful information and interesting patterns is often hidden in these data sets, but it is usually very time-consuming and complicated for us to obtain this information and patterns if relying on the traditionally existing query technology and statistical methods. Therefore, how to automatically and intelligently dig out valuable information and patterns from these data is a challenging problem to solve (Wang et al. 2011). In this case, *data mining* (DM) technology has emerged, which has attracted attention of many scholars worldwide.

Interestingly, it is widely accepted that most of the data in our lives are *geospatial data*, or simply called *spatial data*. Although these data span the data sources available, they are different from transactional data. It is thereby usually difficult to obtain potentially useful knowledge from these spatial data (Li et al. 2013). Li Deren first proposed the concept of "discovering knowledge from GIS database" (KDG) in 1994 (Li et al. 1994). Then he further developed KDG into *spatial data mining* and knowledge discovery (SDMKD) (Li et al. 2002). SDMKD has a wide and important practical role; for example, the rules of spatial association, classification, and clustering are all spatial knowledge that can be applied in many fields (Li et al. 2001).

Spatial data mining is an interesting but very challenging topic not only because of the complexity of spatial autocorrelation, spatial data types, and spatial relations but also because of the continuity of spatial data. Generally, the subsets of spatial features frequently located together in geospatial space are called *spatial co-location patterns* (Yoo and Shekhar 2006; Huang et al. 2004; Yoo et al. 2004, 2005). Co-location pattern mining is a branch of spatial data mining. This pattern explains the association phenomenon in geospatial space, which can provide important information for many application fields. For example, the co-location pattern *{stagnant water source, West Nile disease}* predicts the existence of West Nile disease in stagnant water source areas (Yoo et al. 2006). However, it is not feasible to directly use an association rule algorithm to mine this type of co-location pattern, since the spatial data is different from transactional data, and instances of the spatial features are embedded in a continuous space, sharing multiple spatial relationships (Yoo et al. 2004). The *join-based algorithm*, *partial join algorithm*, and *joinless algorithm* solve the problem of mining co-location patterns and lay a foundation for domestic and foreign scholars to study co-location patterns (Yoo and Shekhar 2006; Huang et al. 2004; Yoo et al. 2004; Zhou 2011; Zhou et al. 2021).

Although co-location pattern mining has been investigated for a few years, most of the co-location mining algorithms have shortcomings. A large part of the

DOI: 10.1201/9781003139416-1

computing time of the join-based algorithm is used to calculate the join to identify the candidate co-location pattern (Huang et al. 2004). Therefore, with the increase of spatial feature types and the number of instances, a large number of join operations are needed, increasing the time-consumption of the join-based algorithm. A partial join algorithm is to establish a group of disjoint clusters between spatial instances to identify the intraX instances and interX instances of co-location, which reduces the number of joins, but it takes time to establish clusters (Yoo et al. 2004). Although a joinless algorithm does not need join operation, it needs to generate table instances and candidate co-location patterns repeatedly, so the efficiency of joinless algorithm is affected by the length of the co-location pattern.

Under this background, the concept of maximal instance and maximal instance algorithm are proposed to mine co-location patterns by Zhou et al. (2021), which overcomes the shortcomings of both these algorithms. With the maximal instance and maximal instance algorithm, the maximal instances in spatial data are found by constructing an instance tree (RI-tree), and the candidate co-location patterns are generated by using the maximal instances. The process does not need any join operation, which greatly reduces the time consumption.

1.2 DATA MINING

1.2.1 Concept for Data Mining

Data mining is the process of automatically extracting hidden useful information from large data repositories in order to find novel and useful patterns that might remain unknown. The data mining has become a powerful technology and tools for (Tan et al. 2006):

- Finding predictive information and patterns, future trends, and behavior that experts may miss,
- Allowing the decision maker to make proactive, knowledge-driven decisions,
- Making prospective analyses and interpretability that are beyond the provision by retrospective tools, such as decision support systems, and
- Answering those questions that traditionally were too laborious and time-consuming to resolve.

Figure 1.1 illustrates a basic architecture of data mining including data collection, selection, transformation, mining and interpretation, and knowledge discovery. The starting point is a data warehouse, where the different types of attribute data and/or spatial data are collected and archived and further managed in a variety of relational database systems. The data mining technology is integrated with the data warehouse to analyze these data using data mining algorithms. The discovered knowledge in the last step is rendered to improve the whole process. Reporting, visualization, and other analysis tools can then be applied to plan the future actions and confirm the impact of those plans.

As seen from Figure 1.1, the process of data mining technology consists of a series of transformation steps, from data preprocessing to post-processing of data

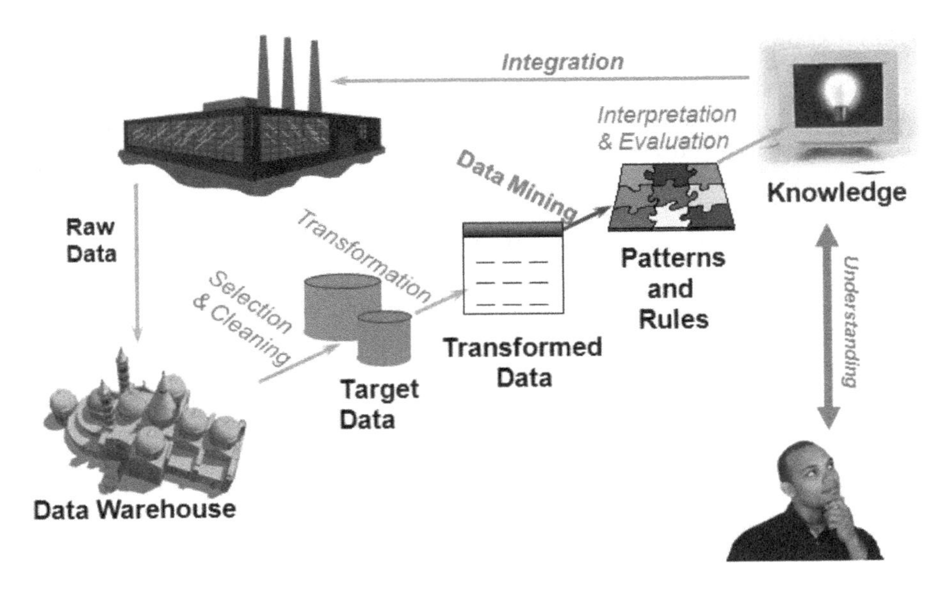

FIGURE 1.1 The basic architecture of data mining.

mining results. As mentioned in Tan et al. (2006), the input data can be in a variety of formats (e.g., flat files, spreadsheets, or relational tables), be stored in a centralized data repository, or be distributed across multiple sites connected by internet. The preprocessing includes fusing data from multiple sources, cleaning data to remove noise and duplicate observations, and selecting records and features that are relevant to the data mining task so that the raw input data can be transformed into an appropriate format for subsequent data mining analysis. The post-processing step is required in order to eliminate spurious data mining results so that only valid and useful results are incorporated into the decision support system. The post-processing algorithm includes statistical measures, hypothesis testing methods, etc. (Tan et al. 2006). Such a "closing the loop" form can ensure the final decision will be as optimal as possible, since the mined information from database can be recycled and refined recursively.

1.2.2 Data Mining and Knowledge Discovery

Knowledge discovery (KD) is a process including data warehousing, target data selection, cleaning, preprocessing, transformation and reduction, data mining, model selection (or combination), evaluation and interpretation, and use of the extracted knowledge (Fayyad 1996). Data mining is an integral part of knowledge discovery in databases (KDD) (Tan et al. 2006). Data mining aims to develop algorithms for extracting new useful patterns from databases that experts may miss, while knowledge discovery aims to enable an information system to transform information to knowledge through hypothesis testing and theory formation (Tan et al. 2006).

1.2.3 Data Mining with Other Disciplines

A number of other disciplines have played key supporting roles in the development of data mining. The germinative idea of data mining was based on sampling, estimation, and hypothesis testing from statistics and search algorithms, modeling techniques, and learning theories from artificial intelligence, pattern recognition, and machine learning (Tan et al. 2006). With the advanced technologies in other disciplines, such as optimization, evolutionary computing, information theory, signal processing, visualization, spatial database, genetic algorithm, and information retrieval, data mining obtained a sustainable development. In particular, techniques from high performance (parallel) computing are often important in addressing the massive size of some data sets, such as database systems for efficient storage, indexing, and query processing. Distributed techniques can also help address the issue of size and are essential when the data cannot be gathered in one location. The most commonly used techniques in data mining are:

- Database technology,
- Information science,
- Statistics,
- Machine learning,
- Visualization, and
- Others such as
  - Artificial neural networks,
  - Decision trees,
  - Genetic algorithms,
  - Classification, and
  - Rule induction.

1.2.4 Data Mining Tasks

Mennis and Guo (2009) have summarized the common tasks in the spatial data mining. These tasks include (1) spatial classification and prediction, (2) spatial association rule mining, (3) spatial clustering regionalization and point pattern analysis, and (4) geo-visualization. Generally speaking, spatial data mining tasks are divided into two major categories (Tan et al. 2006):

- *Predictive tasks.* Like data mining, the major task of spatial data mining is to predict the values and behaviors of attributes on the basis of mined knowledge and patterns.
- *Descriptive tasks.* This task primarily describes the mined knowledge and spatial patterns, such as correlations, trends, neighbors, clusters, trajectories, co-location, co-occurrence, and anomalies.

1.3 GEOSPATIAL DATA MINING

A pavement database is a type of spatial database, since the spatial data, such as XY coordinates, etc., are recorded for describing the pavement location. Thus, study of pavement management on the basis of a pavement database should apply the spatial data mining technology. In fact, spatial data mining is a natural extension of data mining techniques applied to a spatial database. Spatial data mining is also used to extract the useful information and pattern in geography that is unknown and missed by explorers for offering great potential benefits for applied GIS-based decision making. Thus, spatial data mining has the same objectives and goals as data mining does, and even more. Many researchers in information technology (IT), digital mapping, remote sensing, geoinformatics, spatial science, and spatial databases have made tremendous efforts. These efforts include the development of theory, algorithm, methodology, and practice for the extraction of useful information and knowledge from geographically referenced spatial data and drive inductive approaches to geographical analysis and modeling (e.g., Andrienko and Andrienko 1999; Cantú-Paz and Kamath 2000; Chawla et al. 2000; Chen et al. 2005; Guo 2008; Han et al. 1997; Keim et al. 2014; Knorr and Ng 1996; Kulldorff 1997; Mennis and Liu 2010; Miller and Han 2001a, 2001b; Openshaw et al. 1987; Shekhar et al. 2012; Yan and Thill 2009; Yao and Thill 2007; Zhang and Pazner 2004; Huang et al. 2006; May and Savinov 2002; Zhou et al. 2016, 2021; Zhou and Wang 2008, 2010, 2011).

1.4 COMPARISON BETWEEN SPATIAL DATA MINING AND DATA MINING

The common points between spatial data mining and data mining are they can share common methods, algorithms, theories and practices. The differences of the two branches can be briefly summarized as follows (Zhou and Wang 2010, 2011):

1.4.1 Spatial Data in Data Mining

Data describing an object in a spatial database consist of spatial data and nonspatial data. So-called spatial data generally consists of two basic properties – geometric and topological properties. The geometric properties can be spatial location (e.g., geodetic coordinates), area, perimeter, volume, etc. Meanwhile, the topological properties can be adjacency, inclusion, left-/right-hand side, clockwise/counterclockwise, etc. In a traditional database, describing an object usually only uses nonspatial data, that is, no spatial data. The nonspatial data can be stored and managed using a relational database where one attribute of an object has no spatial relationship (Aref and Samet 1991). In a pavement database, the object and event are described by spatial data simultaneously.

In addition, geographic attributes used for describing an object often exhibit the properties of spatial dependency and spatial heterogeneity (Yuan 1997; Gahegan et al. 2001 at www.ucgis.org/priorities/research/research_white/2000%20Papers/emerging/gkd.pdf). The former implies that the attributes at some locations in space are related with others, the latter implies that most geographic processes are unstable, so that global parameters do not well represent the process occurring at a particular location (e.g., Glymour et al. 1997; Han et al. 1993; Hornsby and Egenhofer 2000; Lu et al. 1996; Ng and Han 2002).

These distinct features present challenges and bring opportunities for mining useful information and spatial patterns from nonspatial and/or spatial properties of pavement treatment strategies. Thus, decision tree induction and decision rules induction for pavement management should consider both spatial data and nonspatial data simultaneously. Thus, if the properties of spatial dependency and spatial heterogeneity are ignored, the accuracy of pavement treatment strategies derived from data mining techniques will be affected.

1.4.2 Data Mining and Spatial Database

A pavement database is a type of spatial database. The primary methods for spatial data mining focus on the spatial database, which stores spatial objects represented by spatial data, nonspatial data, and spatial relationships (Han et al. 1993; Agrawal et al. 1993). In addition to extraction of hidden knowledge, spatial patterns, and information, spatial data mining, or knowledge discovery, is also the extraction of implicit spatial relations that are not explicitly stored in spatial databases (Koperski and Han 1995). Also, most studies of spatial data mining focus on the relational and transactional databases. The methods strive to combine already mature techniques such as machine learning, databases, and statistics (Han et al. 1993; Ng and Han 2002).

The fundamental idea of spatial data mining is on the basis of the spatial data of a pavement database, which has some characteristics and brings more challenges than tradition data mining. Existing traditional data mining methods may not have been sufficient to deal effectively with geospatial data, since it can change in spatial and temporal domain. Thus, this research considers the characteristics of spatial data's co-location and co-occurrence.

1.5 DECISION TREES AND DECISION RULES

1.5.1 Decision Tree Induction

Decision tree (DT) induction is one of the most popular and powerful data mining techniques and has thus widely applied in various pattern classifications (Witten and Frank 2002). A decision tree can be understood as a type of classifier that classifies the data set using a tree structure representation of the given decision problem (Osei and Kweku 2007) and is usually composed of three basic elements (Tan et al. 2006) (see Figure 1.2):

1. **A root node,** which is also called *decision node*; it has no incoming edges and zero or more outgoing edges;
2. **Internal nodes**, which is also called *edge*, each of which has exactly one incoming edge and two or more outgoing edges; and
3. **Leaf,** which is also called *terminal node* or *answer node*, each of which has exactly one incoming edge and no outgoing edges.

Over the past few decades, a lot of effort has been made on how to construct an "optimal" decision tree. Dietterich (1990) discussed improvement to decision tree design methods and provided a good background to these and more classical decision tree development methods. Lim et al. (1998) compared several decision trees, such as statistical and neural network methods on a variety of data sets. Both of these works showed that a wide range of speed and accuracies can be obtained from the different decision tree algorithms commonly used, and that the effectiveness of different algorithms varies greatly with the data set. One of the most common "benchmark" methods of inducting decision tree structure is ID3 (Interactive Dichotomizer 3) (Quinlan 1986) and C4.5 (Quinlan 1993), which deals with data sets in which variables are continuous or integer, or where there is missing data, and CART (classification and regression trees) algorithm (Breiman et al. 1984). These algorithms are typically called top-down induction on decision trees (TDIDT), with which the knowledge obtained in the learning process is represented in a tree where each internal node contains a question about one particular attribute (corresponding decision variable) and each leaf is labeled with one of the possible classes (associated with a value of the target variable) (Osei and Kweku 2007). Typical algorithms also include; SLIQ (Mehta et al. 1996), PUBLIC (Rastogi and Shim 2000), SPRINT (Shafer et al. 1996), RAINFOREST (Gehrke et al. 2000), BOAT (Gehrke et al. 1999), MMDT (Chen et al. 2003), and TASC (Chen et al. 2006). In addition, Friedman et al. (1996) discussed the problems of constructing decision trees and showed that the problem of constructing a decision becomes harder as one deals with larger and larger data sets and with more and more variables. Fulton et al. (1996) analyzed the problems of generating decision trees capable of dealing with large, complex data sets and showed that it is simpler to construct decision trees that can deal with a small subset of the original data set. Alsabti et al. (1999) discussed the problems of scaling decision trees up to large data sets, with the loss of accuracy that often occurs as a result. Mehta et al. (1996) emphasized the importance of classification in mining of large data sets and also discussed the wide range of uses that classification can be put to in economic, medical, and scientific situations. Garofalakis et al. (2000) discussed methods for constructing decision trees with user-defined constraints such as size limits or accuracy. These limits are often important for users to be able to understand or use the data sets adequately or to avoid over-fitting the decision tree to the data that is available. Ankerst et al. (1999) used an interactive approach, with the user updating the decision tree through the use of a visualization of the training data. This method resulted in a more intuitive decision tree and one that the user was capable of implementing according to their existing knowledge about the system in question. On the other hand, evolutionary computation for decision tree induction has been of increasing interest to many researchers. Li and Belford (2002) demonstrated that slight changes in the training set could require dramatic changes in the tree topology, that is, the instability inherent in decision tree classifications. Llorà and Garrell (2001) and Papagelis and Kalles (2001) showed that evolutionary methods, when used to develop classification decision trees, allowed both important and unimportant attributes and relationships to be developed and for unimportant factors to be recognized. Cantú-Paz and Kamath (2000), meanwhile, discussed an evolutionary method specifically used to develop classification trees, while Turney (1994) used a definition of fitness for decision tree evolution that included not only error rates but also other costs, such as size. Endou and Zhao (2002) examined a decision tree implementation method that relied on evolution of the training data set used. The training data set was evolved to give the best coverage of the domain knowledge. Siegel (1992) discussed the implementation of competitively evolving decision trees as a method of enhancing evolutionary methods.

Among these methods, one of the most common "benchmark" methods, and also probably the most popular one, is the C4.5 algorithm developed by Quinlan (1986, 1993), which is based on the ID3 method. Thus, this research will emphasize the analysis of the algorithm's advantages and disadvantages in order to present our new method in Chapter 4.

1.5.2 Decision Tree Modeling

In principle, there are exponentially many decision trees that can be constructed from a given set of attributes (Tan et al. 2006), but investigators in fact only endeavor to find a most appropriate decision tree through making a series of locally optimum decisions about which attribute to use for partitioning the data while growing a decision tree, since the optimal tree is computationally infeasible because of the exponential size of the search space (Olaru and Wehenkel 2003). This most appropriate decision tree is believed to be the reasonably most accurate, albeit suboptimal, taking a reasonable amount of time. No matter which algorithm employed, the basic process of a decision tree usually consists of two major phases: the *growth phase* and the *pruning phase* (Apté and Sholom 1997).

1.5.2.1 Growth Phase

The basic process of the growth phase is: a decision tree is generated top-down by successive divisions of the training set in which each division represents a question about an attribute value. The initial state of a decision tree is the root node, which is assigned all of the attributes from the training set. If all attributes belong to the same class, then no further decisions need to be made to partition the attributes, and the solution is complete. If attributes at this node belong to two or more classes, then a split attribute operation will be made by a test. The process is recursively repeated for each of the new intermediate nodes until a completely discriminating tree is obtained (Apté and Sholom 1997). With the generated decision tree, each leaf node is assigned a class label. The nonterminal nodes, which include the root and other internal nodes, contain attribute test conditions to separate records that have different characteristics.

This algorithm, that is, starting from the root to the leaves, is called a *generic decision tree algorithm*, which can be briefly characterized by the following three properties (Elouedi et al. 2001):

1. *Attribute selection measure.* How to choose an attribute is a critical issue because the most appropriate choice will result in partitioning the training set in an *optimized* manner. When a decision node relative to this attribute is created after a test, this node becomes the root of the decision tree.
2. *Partitioning strategy.* How to partition the training set with a given criterion or multiple criteria is very important. It consists in decomposing the training set into many subsets. In order to "optimally" partition the attributes, many

criteria have been presented before; meanwhile, many new criteria are still being proposed.

3. *Stopping criteria.* What criteria will be satisfied for stopping so that a training subset is declared as a leaf? This means that stopping criteria determines whether or not a training subset will be further divided. Some investigators apply the different steps recursively on the training subsets for verifying the stopping criteria.

One of the most important properties is the attribute selection measurement, which measures how to select the attribute that characterizes the root of the decision tree and those of the different subdecision trees. Quinlan (1993) has defined a measure called *information gain* and further developed a well-known popular decision tree modeling algorithm called *C4.5*. The details of attribute selection measures will be described in section 1.5.3. Briefly, the basic idea of this attribute selection measure is to compute the information gain of each attribute in order to find how well each attribute alone classifies the training examples, and then one presenting the highest value will be chosen. In fact, this attribute generates a partition where the record classes are as homogeneous as possible within each subset created by the attribute.

In order to explain this basic process, Figure 1.2 illustrates the data set, and the corresponding tuple-table is listed in Table 1.1. The figure also shows the three axis parallel lines, one at height = 4.2, the second line at height = 6.3, and third line at volume = 34. The three lines seem to completely partition the training data set into three different subareas.

Figure 1.2 illustrates the process of a decision tree growth phase for training set listed in Table 1.1 and Figure 1.2. In the first step, all attributes are assigned to the

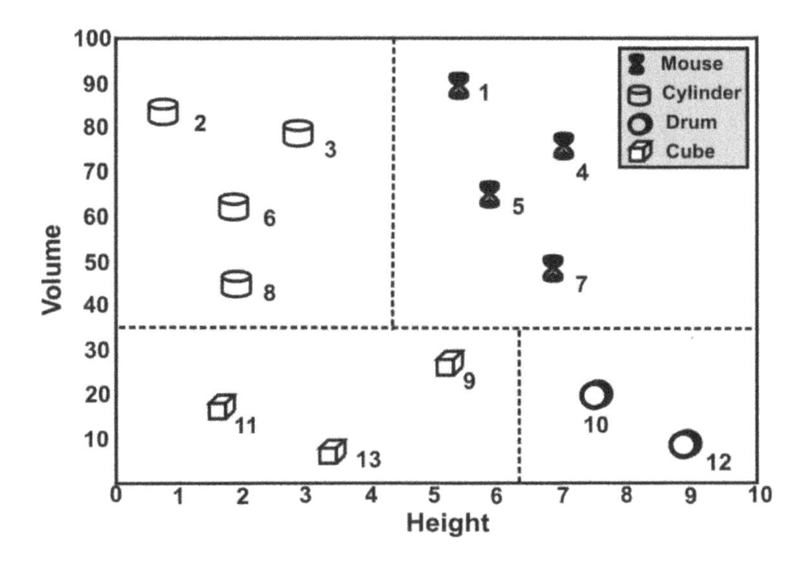

FIGURE 1.2 Example of data set for decision tree generation.

TABLE 1.1
Example of Data Set for Decision Tree Generation

ID	Height	Volume	Class
1	5.2	91.3	
2	0.85	84.4	
3	2.96	78.3	
4	6.99	75.5	
5	5.92	65.0	
6	1.87	62.2	
7	6.83	48.5	
8	1.79	45.6	
9	5.33	26.4	
10	7.55	19.2	
11	1.87	18.4	
12	8.99	8.9	
13	3.41	8.2	

top level of the tree, that is, root node, at which the classification process begins with a condition test for all examples at *volume* > 34. Examples that satisfy this test conditions with TRUE are passed down to the left internal node, with FALSE are passed down to the right internal node. This means that the right edge from root node receives examples that are not yet purely from one class, so further testing is required at this intermediate node. The second test at this level for the left node (TRUE) is for *height* > 4.2, and for the right node (FALSE) is for *height* > 6.2. For the left node, examples that satisfy the test condition (*height* > 4.2) are all in one class (MOUSE), and those that do not are all in another class (CYLINDER). For the right node, examples that satisfy that test condition (*height* > 6.2) are all in one class (DRUM), and those that do not are all in another class (CUBE). At this stage, both edges from this node lead to leaf nodes, that is, no more tests are needed; thus, the decision tree solution is complete. Note that this example illustrates a binary tree, where each intermediate node can split into at most two sub-trees. In fact, a decision tree may be nonbinary tree, where each intermediate node may split into more than two sub-trees.

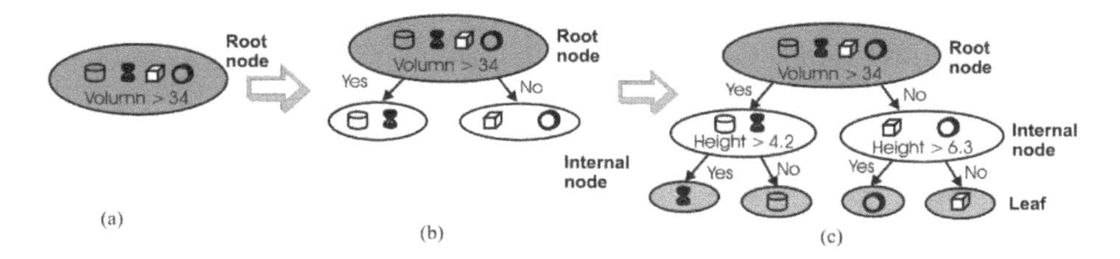

FIGURE 1.3 Process of tree growth phase.

1.5.2.2 Pruning Phase

Due to noise and outliers in the training data, the generated decision tree at the previous stage is potentially an over-fitted solution. The over-fitting can heavily influence the classification accuracy of new data sets. Thus, a second phase, called *pruning*, is required to eliminate sub-trees in order to minimize the real misclassification error produced in growth phase (Apté and Sholom 1997). The actions of the *pruning phase* are often referred to as *post-pruning* in contrast to the *pre-pruning* that occurs during the *growth phase*. In order to create a small and interpretable decision tree, numerous post-pruning methods have been proposed (e.g. Almuallim 1996; Bohanec and Bratko 1994; Fournier and Cremilleux 2002; Li et al. 2001; Mingers 1989, 1987; Niblet and Bratko 1986; Quinlan 1986, 1987, 1993; Mansour 1997; Elouedi et al. 2001; Säuberlich 2000; Witten and Frank 2002). These methods can be grouped by (Osei and Kweku 2007):

- *Error-based method.* Some post-pruning approaches attempted to identify a sub-tree that gives the smallest error on the validation data set, such as the reduced error pruning method proposed by Quinlan (1987), while others use an error estimation that is derived from training data set only, such as the minimum error pruning method developed by Niblet and Bratko (1986).
- *Top-down or down-top method.* Some researchers propose a top-down approach, such as the pessimistic error method (Quinlan 1987); while some researchers take a bottom-up approach, such as the error-based pruning method (Quinlan 1993).
- *Optimal or suboptimal method.* Some methods are suboptimal heuristics (e.g. Mingers 1987); some methods are proposed to produce optimal solutions (e.g. Almuallim 1996; Bohanec and Bratko 1994).
- *Criterion method.* Some methods used a signal criterion (e.g., Quinlan 1987); some methods used a multi-criteria approach for evaluating the "best" DT in a set of generated DTs (e.g., Osei and Kweku 2007, 2004).

1.5.3 Measures for Selecting the Best Split

Many measures have been developed to determine the best way to split the attributes based on the degree of impurity of the child nodes during the growth phase of a decision tree. Most of these measures are defined in terms of the class distribution of the records before and after splitting. The smaller the degree of impurity, the more skewed the class distribution (Tan et al. 2006). The commonly used standard splitting measures are entropy (Quinlan 1986), gain ratio (Quinlan 1993) and Gini index (Breiman et al. 1984). The first two measures will be used in this research.

1.5.3.1 Entropy

In information theory, entropy is a measure of the uncertainty associated with a random variable. Also, the entropy is a measure of the average information content one is missing when one does not know the value of the random variable (http://en.wikipedia.org/wiki/Entropy (information _theory)). Entropy was first adopted in decision tree generation by Quinlan (1986) in his ID3 algorithm as a split measure. The formula is (Tan et al. 2006)

$$Entropy(t) = -\sum_{i=0}^{c-1} p(i\,|\,t)\log_2 p(i\,|\,t) \tag{1.1}$$

Where $p(i\,|\,t)$ is the fraction of records belonging to class i at a given node t, and c is the number of classes. The ID3 algorithm utilized the entropy criteria for splitting nodes. The process is: Giving a node t, computing the splitting criterion, $Entropy\ (t) = p_i \times \log(p_i)$, where p_i is the probability of class i within node t. An attribute and split are selected that minimize entropy. Splitting a node produces two or more direct descendants. Each child has a measure of entropy. The sum of each child's entropy is weighted by its percentage of the parent's cases in computing the final weighted entropy used to decide the best split.

1.5.3.2 Information Gain

For a training set T on attribute A, information gain in information theory and machine learning is defined as (Elouedi et al. 2001):

$$Gain(T, A) = Info(T) - Info_A(T) \tag{1.2}$$

where

$$Info(T) = \sum_{i=1}^{n} \frac{freq(C_i, T)}{|T|} \cdot \log_2 \frac{freq(C_i, T)}{|T|} \tag{1.3}$$

$$Info_A(T) = \sum_{v \in D(A)}^{n} \frac{|T_v|}{|T|} \cdot Info(T_v) \tag{1.4}$$

where $\Theta = \{C_1, C_1, \cdots, C_n\}$ are the set of n mutually exclusive and exhaustive classes, $freq(C_i, T)$ denotes the number of objects in the set T that belong to the class C_i, and T_v is the subset of objects for which that attribute A has the value v.

Theoretically, the best attribute is the one that maximizes *Gain* (T, A). Once the best attribute is allocated to a node, the training set T is split into several subsets. The procedure is then iterated for each subset.

1.5.3.3 Gain Information Ratio

Elouedi et al. (2001) demonstrated that the gain information has good results, but it is limited to those attributes with a large number of values over those with a small number of values. To overcome this drawback, Quinlan (1993) has proposed the gain ratio criterion, which is mathematically defined by:

$$Gain\ ratio(T,A) = \frac{Gain(T,A)}{Split\ Info(T,A)} \tag{1.5}$$

where $Split\ Info(T,A) = -\sum_{v \in D(A)} \frac{|T_v|}{|T|} \cdot \log_2 \frac{|T_v|}{|T|}$ measures the information in the attribute due to the partition of the training set T into $|D(A)|$ training subsets. Split info (T, A) is also the information due to the split of S on the basis of the value of the categorical attribute A. With gain ratio, the attributes with many values will be adjusted.

In C4.5 algorithm (Quinlan 1993), the attribute value that maximizes the gain ratio is chosen for the splitting attribute. The gain ratio is computed using attributes having gain greater than average gain. This gain ratio expresses the proportion of information generated by a split that is helpful for developing the classification. The numerator (the information gain) in this ratio is the standard information entropy difference achieved at node t, expressed in Equation 1.4.

1.5.4 Decision Rule Induction

Decision rules are directly induced by translating a decision tree either in a bottom-up specific-to-general style or in a top-down general-to-specific style (Apté and Sholom 1997). In other words, the decision rules are constructed by forming a conjunct of every test that occurs on a path between the root node and a leaf node of a tree.

Algorithms of inducing decision rules can be grouped into two categories: ordered rule sets or unordered rule sets (Apté and Sholom 1997):

1. Ordered rule sets are induced by ordering all the classifications, and then using a fixed sequence, such as the smallest to the largest class, to combine them together. When this rule is applied to new data set, the new data example is required in exactly the same sequence as they were generated in the training data. Based on the example in Figure 1.3, the induced decision rules are depicted in Figure 1.4.
2. Unordered rule sets are induced without a fixed sequence. Thus, when this rule is applied to new data, the new data example can be independent and more flexible.

For the basic process of decision rule induction, that is, a tree generation first and then translation of the tree into a set of rules, some problems were discovered. For example, for certain data spaces, this nature of partitioning may not always be capable of producing appropriate/optimal solutions. On the other hand, if algorithms

```
IF (volume > 34)
THEN If (height > 4.2)
    Then MOUSE
    ELSE CYLINDER
ELSE IF
  IF (height > 6.3)
    THEN DRUM
    ELSE CUBE
End
```

FIGURE 1.4 Decision rule induction.

are employed that directly generate the tree, it is possible to create rules. These rules essentially correspond to decision regions that overlap each other in the data space. Thus, some people suggested the techniques that directly generate rules from data are also available, which overcome some of the drawbacks of decision tree modeling.

1.5.5 Evaluation of the Performance of Decision Trees

Once a decision tree and/or decision rule is induced, it can be used for estimating or predicting a new data set. Many methods have been developed to evaluate the performance of a decision tree or decision rules. The most well-known criteria are accuracy, speed, and interpretability. In other words, the decision tree and decision rules derived using different approaches can be compared in terms of their predictive accuracy on the new data set, on the computational cost, and the level of understanding and insight that is provided by the solution. Accuracy and speed vary from algorithm to algorithm, and in most instances these two issues are coupled; that is, a high predictive accuracy tends to require increased computational effort (Apté and Sholom 1997). This research will use the following criteria to evaluate the performance of a decision tree and decision rules.

1.5.5.1 Accuracy of Performance

The performance accuracy of a decision tree is defined as a ratio between the number of correct or incorrect classified instances. The mathematical formula is (Tan et al. 2006):

$$Accuracy = \frac{Number\ of\ correct\ predictions}{Total\ number\ of\ predictions} \tag{1.6}$$

This classification of accuracy gives a general assessment of the number of correctly classified examples in total.

1.5.5.2 Twofold Cross-Validation

Twofold cross-validation will be applied in this research to evaluate the performance of decision tree and decision rules. The basic process is: the whole data set is split into two parts, one part of the data set being dedicated to the training and the other one for the test. The training set is used to learn the algorithm and generate the tree and rules, and the test set is used to estimate the generated decision tree and rules. This procedure is repeated after every part of the data set is used for both training and testing, respectively. Afterward, the overall accuracy parameters are calculated as means from the evaluation of the individual cross-validation subset.

1.5.6 Problems of Decision Tree Induction Data Mining

Decision tree induction is capable of extracting implicit, previously unknown, and potentially useful information from large databases and has therefore been successfully and widely used in various domains, including data mining (Quinlan 1986, 1993) and many other industrial and business domains for credit evaluations, fraud detection, and customer-relationship management (Berry and Linoff 2000).

The decision tree induction method has several advantages over other data mining methods, including being human interpretable, well organized, computationally inexpensive, and capable of dealing with noisy data (Li et al. 2001).

However, the decision tree induction method entails the following drawbacks:

1. Up to until now, decision tree construction algorithms have usually assumed that the class labels were Boolean variables. This means that the algorithms operate under the assumption that the class labels are flat. In other words, decision tree construction takes each attribute through in a one-by-one manner without considering the simultaneous occurrence of multiple attributes. In real-world applications, there are more complex class scenarios, where the classification labels to be predicted are in co-occurrence labels. Unfortunately, existing research has paid little attention to the classification of data with co-occurrence class labels. To the best of our knowledge, no method has been developed to construct DTs directly from data are in co-occurrence class labels. This research work intends to remedy this research gap.
2. Almost all of the decision tree generation methods did not consider the spatial features of geospatial data, such as geographic relationships and topological relationships. In other words, the spatial data contains objects that are characterized by a spatial location and/or extension as well as by several nonspatial attributes. Figure 1.5 shows an example of spatial objects, which occur at a co-location pattern, that is, CYLINDER always co-occurs with MOUSE. In a real-world, some instances are often located close geographically to another instance, such as gasoline station and road. Thus, identification of such a classification pattern associated with spatial relationships and topological relationships needs to be studied.

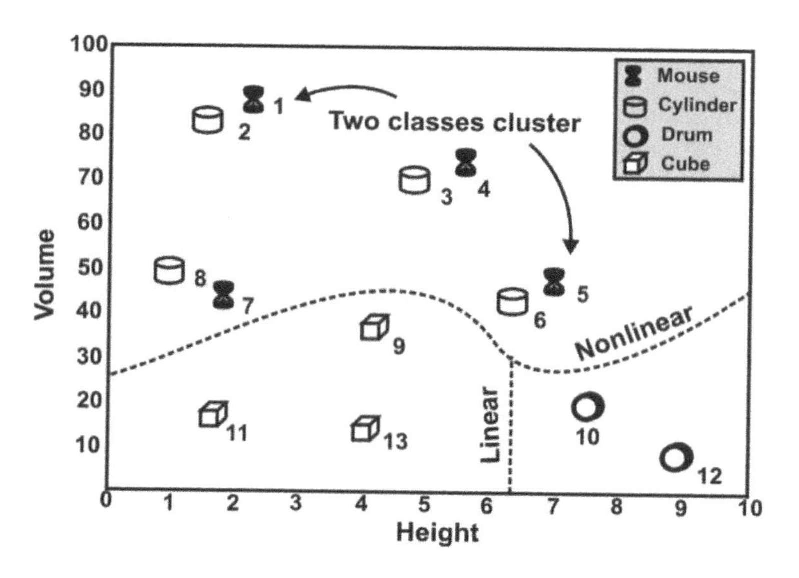

FIGURE 1.5 Instance co-location pattern and nonlinear classification.

3. Mugambi et al. (2004) divided the decision trees into three main types on the basis of how they partition the feature space:

    - *Univariate or axis-parallel decision tree.* This type of decision tree carries out tests on a single variable at each non-leaf node and splits the attributes using axis-parallel hyperplanes in the feature space (see Figure 1.2). The C4.5 algorithm (Quinlan 1993) belongs to the axis-parallel class of decision trees. This type of tree is called a *linear decision tree.*
    - *Multivariate linear or oblique.* This type of decision tree carries out tests and split the attributes using an oblique orientation to the axis of the feature space geometrically.
    - *Nonlinear multivariate decision trees.* This type of decision trees carries out tests using nonlinear partitioning of the feature space (see Figure 1.5), such as polynomial-fuzzy decision tree (Mugambi et al. 2004).

A linear decision tree is known to perform well in small and linear feature spaces but very poorly in large and nonlinear ones. Theoretically, exploring information patterns using decision tree is based on a large database. In fact, in our pavement management database, the database is not large enough, as expected in principle. This means that the pavement data mining uses linear decisions better than a nonlinear decision tree. However, the spatial features in the pavement database are not a linear mode in the real-world. Thus, this fact requires us to develop a robust linear decision tree method to handle small data with linear spatial features.

1.6 CO-LOCATION PATTERN MINING

Many scholars have studied co-location patterns earlier and achieved many satisficing research results. The use of the co-location pattern is proposed by Huang et al. (2004). The basic mining algorithms for a co-location pattern are join-based algorithms, partial join algorithms, and joinless algorithms. Literature (Huang et al. 2004) not only introduces the concept of the co-location pattern but also proposes a join-based algorithm to solve the problem of co-location pattern mining. The join-based algorithm is also called a transaction-free algorithm, because there is not a customized transaction set in the spatial data set. This algorithm uses the principle of data mining to discover the co-location pattern in spatial data without the transaction set. In the join-based algorithm, the participation index satisfies an anti-monotonic property, which not only provides an effective pruning strategy to reduce unnecessary computing time but also ensures the correctness of co-location pattern mining.

Partial join algorithm was first proposed in the literature by Yoo and Shekhar (2004). This algorithm is to solve the problem of poor performance caused by too many join operations in the join-based algorithm. The main idea is to reduce the calculation time by reducing the join operations, so as to achieve the purpose of optimizing the algorithm. Before mining the co-location pattern, the partial join algorithm first transacts the spatial data set, that is, generating disjoint clusters in the data set.

The joinless algorithm was first proposed by Yoo and Shekhar (2004). This algorithm is to better solve the problem of excessive computation consumption caused by join operation in the join-based algorithm and partial join algorithm. The main idea of a joinless algorithm is to establish a star neighborhood in a spatial data set. Compared with the join-based algorithm and partial join algorithm, this algorithm is more effective because it uses the instance lookup method rather than the computationally expensive instance join method when obtaining instances of the co-location pattern.

Co-location pattern studies have tended to emphasize the equal participation of each spatial feature. As a result, it is impossible to capture interesting patterns involving features with different frequencies. Therefore, Huang, Pei, and Xiong studied the mining of co-location patterns with rare spatial features and proposed a new measure called maximum participation ratio (maxPR) (Ma 2017).

In addition, a few scholars have put forward the concept of maximal clique in co-location pattern mining, which can be used to mine co-location patterns quickly and effectively. For example, Verhein and Al-Naymat (2007) first proposed the concept of maximal clique; Verherin et al. (2007) also introduced the data of the Sloan Digital Sky Survey (SDSS) and proposed an effective algorithm to generate the maximal clique from a large spatial database; this algorithm can successfully generate all the maximal cliques in SDSS data and generate a useful co-location pattern; Kim et al. (2011) proposed a polynomial algorithm called AGSMC (Algorithm Generating Spatial Maximal Cliques), which can generate all the maximal cliques from the general spatial data sets (Kim et al. 2011). This algorithm uses the materialized method to construct a tree data structure to represent the maximal cliques, which can effectively mine the spatial co-location pattern (Ghazi-Tabatabai et al. 2008).

A few other scholars have studied the relationship between spatial co-location pattern and spatial clustering. For example, Huang et al. (2008) proposed spatial clustering to combine similar spatial objects and introduced a new method of mining spatial co-location patterns by clustering technology; Jiamthapthaksin (2009) proposed multi-objective clustering (MOC), which decomposes data sets into similar groups and maximizes multiple objects in parallel, and applied the MOC to co-location pattern mining (Huang et al. 2008); Yu (2017) proposed a new co-location analysis method to analyze the popular areas of the patterns; this method combines kernel density estimation and polygon clustering technology and specifically considers the correlation, heterogeneity, and context information in complex spatial interaction (Jiamthapthaksin et al. 2009). Bian and Wan (2009) defined a spatial co-location pattern based on k-proximity in reference(Wu et al. 2013) and proposed a k-proximity feature co-location pattern mining algorithm based on grid index to effectively discover co-location patterns in spatial data sets. Wu et al. (2013) proposed the basic algorithm of fuzzy co-location mining based on the related concepts of fuzzy co-location mining. Jiang et al. (2017) introduced the spatial instance with utility value and the new utility participation index and used the new utility participation index to mine co-location patterns (Gao et al. 2011). Yang et al. (2014) proposed an agglomerative hierarchical clustering algorithm for spatial co-location pattern mining. Hu et al. (2014) introduced the concept of utility into spatial co-location pattern mining and proposed a basic algorithm to mine spatial co-location pattern with high efficiency; Hu et al. (2013) studied spatial co-location pattern mining with instance location ambiguity, defined the relevant concepts of spatial co-location pattern mining with instance location ambiguity, and proposed the distance calculation based on grid and pruning strategy to improve the mining performance and accelerate the generation of co-location rules. The concept of spatial maximal co-location pattern and the mining algorithm were first proposed by Lu et al. (2014). Guo et al. (2016) proposed the degree method according to the characteristics of large degree of the maximal clique vertex and then studied the efficient incremental mining of spatial co-location pattern and the evolution analysis of spatial co-location pattern and proposed the basic algorithm and pruning algorithm of efficient incremental mining of spatial co-location patterns.

With the development of spatial data mining technology, the research interests on the co-location pattern are increasing, and this pattern has been widely used in various fields. Zhou (2011) applied the co-location pattern to the decision tree and proposed a decision tree induction method called *co-location based decision tree (CL-DT)* to strengthen the decision of pavement maintenance and repair. In addition, since there is a nonlinear distribution of instances in high-dimensional space, the CL-DT decision tree induction only considers the Euclidean distance between instances. He (2018) and Jiang et al. (2010) used co-location patterns to analyze the urban planning and supermarket distribution. In order to make the analysis and research of co-location more practical, Hu et al. (2008) combines the analysis of co-location patterns with the spatial analysis method of a geographic information system, put forward the theory of co-location spatial analysis based on buffers, and studied the influencing factors of karst collapse on this basis. The concept of spatiotemporal co-location patterns was given by He (2018), who studied urban traffic data and used spatiotemporal co-location pattern mining to find the collection of congestion and transitive roads in urban traffic, which provides good suggestions for solving the problem of urban congestion.

1.7 ARRANGEMENT OF THE CHAPTERS

Several of the apparent features of the book that I believe will enhance its value as a textbook and as a reference book are as follows: First, this book only concentrates on a special topic, mining co-location patterns; that is, this book goes beyond the fundamentals of data mining, such as data analysis, pattern mining, clustering, classification, statistics, regression, neural networks, deep learning, and machine learning. In fact, no single book has addressed all these topics in a comprehensive and integrated way. Second, co-location pattern mining is an emerging field and a cutting-edge topic in data science and is able to automatically dig co-location patterns from all types of data sets, with applications ranging from scientific discovery to business intelligence and analytics. Third, this book emphasizes the principles, algorithms, and wide applications of co-location pattern mining; this book can therefore advance readers' knowledge and skill in data mining. Fourth, the content of this book is specially deliberated with the following arrangements. Such a deliberated chapter arrangement is to help readers completely understand the principle of mining co-location patterns step by step.

Chapter 1: Introduction. The background and significance of this book, fundamentals of data mining, and the advance of co-location patterns are overviewed and introduced.

Chapter 2: Fundamentals of mining spatial co-location patterns. The definitions and concepts related to mining co-location patterns are introduced. The algorithms for three types of mining co-location patterns, including join-based, partial join-based, and joinless-based, are introduced. Their advantages and disadvantages are discussed, and some of the other algorithms developed in recent years are reviewed in detail.

Chapter 3: Principles of mining spatial co-location patterns. Methods and algorithms for mining co-location patterns are explained in detail. The generation of co-location decision trees and decision rule induction from co-location are described.

Chapter 4: Manifold learning co-location pattern mining. Manifold learning is first introduced, and the maximum variance unfolding (MVU)-based mining co-location pattern, including generation of rule and pruning, is focused on and detailed.

Chapter 5: Maximal instance algorithm for mining co-location patterns. The maximal instance algorithm, through introducing a new concept, maximum instance, is described in detail. The process of generating row instances and co-location patterns in this algorithm is introduced. Its advantages and disadvantages are discussed.

Chapter 6: Mining negative co-location patterns. S few new definitions and concepts pertaining to negative co-location are introduced. The principle of mining negative co-location patterns is described in detail. The relationship between positive and negative co-location patterns is analyzed.

Chapter 7: Applications of mining co-location patterns in pavement maintenance and rehabilitation are described in detail. Comparison analysis and discussion are introduction, and decision making is recommended.

Chapter 8: The application of mining co-location in spatial buffer analysis is described in detail, including three traditional types of buffer analysis, hat is, point buffering, line buffering, and polygon buffering. A comparison analysis and remarks are made.

Chapter 9: The application of mining co-location patterns in remote sensed imagery classification is introduced in detail. A comparison analysis with the different methods is discussed.

REFERENCES

Agrawal, R., Imielinski, T., and Swami, A., Mining association rules between sets of items in large databases. *Proc. 1993 ACM-SIGMOD Int. Conf. Management of Data*, Washington, DC, May 1993, pp. 207–216.

Almuallim, H., An algorithm for the optimal pruning of decision trees. *Artificial Intelligence*, vol. 83, no. 2, 1996, pp. 347–362.

Al-Naymat, G., Enumeration of maximal clique for mining spatial co-location patterns. *IEEE/ACS International Conference on Computer Systems & Applications*, Doha, Qatar, March 31–April 4, 2008, pp. 126–133.

Alsabti, K., Ranka, S., and Singh, V., Clouds: A decision tree classifier for large datasets. *Proceedings of the Fourth International Conference on Knowledge Discovery and Data Mining (KDD-98)*, AAAI Press, New York, USA, August 27–31, 1999, pp. 2–8.

Andrienko, G., and Andrienko, N., Data mining with C4.5 and interactive cartographic visualization. In N.W.G.T. Paton (Ed.), *User interfaces to data intensive systems*. Los Alamitos, CA: IEEE Computer Society, 1999, pp. 162–165.

Ankerst, M., Elsen, C., Ester, M., and Kriegel, H.P., Visual classification: An interactive approach to decision tree construction. *Proceedings of International Conference on Knowledge Discovery and Data Mining (KDD'99)*, San Diego, CA, August 15–18, 1999, pp. 392–396.

Apté, C., and Sholom, W., Data mining with decision trees and decision rules. *Future Generation Computer Systems*, vol. 13, no. 2–3, November 1997, pp. 197–210.

Aref, W.G., and Samet, H., Optimization strategies for spatial query processing. *Proc. 17th Int. Conf. VLDB*, Barcelona, Spain, September 1991, pp. 81–90.

Berry, M., and Linoff, G., *Mastering data mining: The art and science of customer relationship management*. New York: John Wiley & Sons, Inc, 2000.

Bian, F.L., and Wan, Y., A novel spatial co-location pattern mining algorithm based on k-nearest feature relationship. *Geomatics and Information Science of Wuhan University*, vol. 34, no. 3, 2009, pp. 331–334.

Bohanec, M., and Bratko, I., Trading accuracy for simplicity in decision trees. *Machine Learning*, vol. 15, 1994, pp. 223–250.

Breiman, L., Friedman, J.H., Olshen, R.A., and Stone, C.J., *Classification and regression trees*. Belmont, CA: Wadsworth, The 1st Version (Jan 1, 1984), Chapman and Hall/CRC, ISBN-13: 978–0412048418.

Cantú-Paz, E., and Kamath, C., Using evolutionary algorithms to induce oblique decision trees. In D. Whitley, D.E. Goldberg, E. Cantú-Paz, L. Spector, I. Parmee, and H.-G.

Beyer (Eds.), *GECCO-2000: Proceedings of the genetic and evolutionary computation conference*. San Francisco, CA: Morgan Kaufmann, 2000, pp. 1053–1060.

Chawla, S., Shekhar, S., Wu, W., and Ozesmi, U., Extending data mining for spatial applications: A case study in predicting nest locations. *ACM SIGMOD Workshop on Research Issues in Data Mining and Knowledge Discovery (DMKD 2000)*, Dallas, TX, May 14, 2000, pp. TR 00-026-1-12.

Chen, Y.L., Hsu, C.L., and Chou, S.C., Constructing a multivalued and multi-labeled decision tree. *Expert Systems with Applications*, vol. 25, no. 2, 2003, pp. 199–209.

Chen, Y.L., Hsu, W.H., and Lee, Y.H., TASC: Two-attribute-set clustering through decision tree construction. *European Journal of Operational Research*, vol. 174, no. 2, 2006, pp. 930–944.

Chen, Y.L., Tang, K., Shen, R.J., and Hu, Y.H., Market basket analysis in a multiple store environment. *Decision Support Systems*, vol. 40, no. 2, 2005, pp. 339–354.

Dietterich, T.G., Machine learning. *Annual Review of Computer Science*, vol. 4, 1990, pp. 255–306.

Elouedi, Z., Khaled, M., and Philippe, S., Belief decision trees: Theoretical foundations. *International Journal of Approximate Reasoning*, vol. 28, no. 2–3, November 2001, pp. 91–124.

Endou, T., and Zhao, Q.F., Generation of comprehensible decision trees through evolution of training data. *Proceedings of IEEE Congress on Evolutionary Computation (CEC'2002)*, Honolulu, HI, pp. 1221–1225.

Fayyad, U., Piatetsky-Shapiro, G., and Smyth, P., From data mining to knowledge discovery in databases. *AI Magazine*, Fall 1996, pp. 37–54.

Fournier, D., and Cremilleux, B., A quality index for decision tree pruning. *Knowledge-Based Systems*, vol. 15, 2002, pp. 37–43.

Friedman, J.H., Kohavi, R., and Yun, Y., Lazy decision trees. *Proceedings of the 13th National Conference on Artificial Intelligence and Eighth Innovative Applications of Artificial Intelligence Conference*, vol. 1. AAAI Press/The MIT Press. AAAI 96, IAAI 96, August 4–8, 1996, pp. 717–724.

Fulton, T., Kasif, S., Salzberg, S., and Waltz, D., Local induction of decision trees: Towards interactive data mining. *Proceedings of Second International Conference on Knowledge Discovery and Data Mining*, Portland, OR, August 1996, pp. 14–19.

Gahegan, M., Wachowicz, M., Harrower, M., and Rhyne, T.M., The integration of geographic visualization with knowledge discovery in databases and geocomputation. *Cartography and Geographic Information Systems*, special issue on the ICA research agenda. *Cartography and Geographic Information Science*, vol. 28, no. 1, 2001, pp. 29–44.

Gao, S.J., Wang, L.Z., Feng, L., et al., Co-location patterns mining based on agglomerative hierarchical clustering. *Journal of Guangxi Normal University: Natural Science Edition*, vol. 29, no. 2, 2011, pp. 167–173.

Garofalakis, M., Hyun, D., Rastogi, R., and Shim, K., Efficient algorithms for constructing decision trees with constraints. *Proceedings of the 6th ACM SIGKDD International Conference on Knowledge Discovery and Data Mining*, Boston, MA, August 2000, pp. 335–339.

Gehrke, J., Ganti, V., Ramakrishnan, R., and Loh, W.-Y., BOAT-optimistic decision tree construction. *Proceedings of the 1999 ACM SIGMOD International Conference on Management of Data*, Philadelphia, PA, 1999, pp. 169–180.

Gehrke, J., Ramakrishnan, R., and Ganti, V., RainForest: A framework for fast decision tree construction of large datasets. *Data Mining and Knowledge Discovery*, vol. 4, no. 2–3, 2000, pp. 127–162.

Ghazi-Tabatabai et al., Structure and disassembly of filaments formed by the ESCRT-III subunit Vps24. *Structure*, vol. 16, no. 9, September 10, 2008, pp. 1345–1356.

Glymour, C., Madigan, D., Pregibon, D., and Smyth, P., Statistical themes and lessons for data mining. *Journal of Data Mining and Knowledge Discovery*, vol. 1, 1997, pp. 11–28.

Guo, D., Regionalization with Dynamically Constrained Agglomerative Clustering and Partitioning (REDCAP). *International Journal of Geographical Information Science*, vol. 22, no. 7, 2008, pp. 801–823.

Guo, Y.M., Liu, Y., Oerlemans, A., Lao S.Y., Wu, S., and Lew, M.S., Deep learning for visual understanding: A review. *Neurocomputing*, vol. 187, 2016, pp. 27–48.

Han, J., Cai, Y., and Cercone, N., Data-driven discovery of quantitative rules in relational databases. *IEEE Transaction Knowledge and Data Engineering*, vol. 5, 1993, pp. 29–40.

Han, J.W., Koperski, K., and Stefanovic, Geominer: A system prototype for spatial data mining. *Proceedings of the 1997 ACM SIGMOD international conference on Management of data,* Tucson, Arizona, USA,1997, pp. 553–556.

He, Y., *Spatio-temporal co-location pattern mining and applications in urban congestion identification.* Kunming: Yunnan University, 2018.

Hornsby, K., and Egenhofer, M.J., Identity-based change: A foundation for spatiotemporal knowledge representation. *International Journal of Geographical Information Science*, vol. 14, 2000, pp. 207–224.

Hu, C.P., and Qin, X.L., A novel positive and negative spatial co-location rules mining algorithm. *Journal of Chinese Computer Systems*, vol. 29, no. 1, 2008, pp. 80–84.

Hu, X., Wang, L.Z., He, W., et al., Degree-based approach for the maximum clique problem. *Journal of Frontiers of Computer Science and Technology*, vol. 7, no. 3, 2013, pp. 262–271.

Hu, X., Wang, L.Z., Zhou, L.M., et al., Spatial maximal co-location patterns. *Journal of Frontiers of Computer Science and Technology*, no. 2, 2014, pp. 26–36.

Huang, Y., Pei, J., and Xiong, H., Mining co-location patterns with rare events from spatial data sets. *Geoinformatica, Durkin*, vol. 10, no. 3, 2006, pp. 239–260.

Huang, Y., Shekhar, S., and Xiong, H., Discovering colocation patterns from spatial data sets: A general approach. *IEEE Transactions on Knowledge and Data Engineering*, vol. 16, no. 12, 2004, pp. 1472–1485.

Huang, Y., Zhang, P.S., and Zhang, C.Y., On the relationships between clustering and spatial co-location pattern mining. *International Journal on Artificial Intelligence Tools*, vol. 17, no. 1, 2008, pp. 55–70.

Jiamthapthaksin, R., Eick, C.F., and Vilalta, R., A framework for multi-objective clustering and its application to co-location mining. In R. Huang, Q. Yang, J. Pei, J. Gama, X. Meng, and X. Li (Eds.), *Advanced data mining and applications.* ADMA. Lecture Notes in Computer Science, vol. 5678. Berlin, Heidelberg: Springer, 2009.

Jiang, W.G., Wang, L.Z., Fang, Y., et al., Domain-driven high utility co-location pattern mining method. *Journal of Computer Applications*, no. 2, 2017, pp. 322–328.

Jiang, Y., Wang, L., Lu, Y., et al., Discovering both positive and negative co-location rules from spatial data sets. *International Conference on Software Engineering & Data Mining*, IEEE, Chengdu China, June 28–30, 2010, pp. 241–250.

Keim, D.A., Panse, C., Sips, M., and North, S.C., Visual data mining in large geospatial point sets. *IEEE Computer Graphics and Applications*, vol. 24, no. 5, 2014, pp. 36–44.

Kim, K.S., Kim, Y.H., and Kim, U., Maximal cliques generating algorithm for spatial colocation pattern mining. *FTRA International Conference on Secure & Trust Computing*, Loutraki, Greece, June 28–30, 2011, pp. 241–250.

Knorr, E.M., and Ng, R.T. Finding aggregate proximity relationships and commonalities in spatial data mining. *IEEE Transactions on Knowledge and Data Engineering*, vol. 8, no. 6, 1996, pp. 884–897.

Koperski, K., and Han, J., Discovery of spatial association rules in geographic information databases. *Proc. 4th International Symposium on Large Spatial Databases*, SSD95, Maine, 1995, pp. 47–66.

Kulldorff, M., A spatial scan statistic. *Communications in Statistics-Theory and Methods*, vol. 26, 1997, pp. 1481–1496.

Li, D., and Cheng, T., KDG-knowledge discovery from GIS: Propositions on the use of KDD in an Intelligent GIS. *The Canadian Conference on GIS*, Ottawa, Canada, June 6–10, 1994, pp. 1001–1012.

Li, D., and Li, D.Y., Theories and technology of spatial data mining and knowledge discovery. *Geomatics and Information Science of Wuhan University*, vol. 27, no. 3, 2002, pp. 221–233.

Li, D., Wang, S., and Li, D.W., *Spatial data mining theory and applications*. Beinjing, China: Academic Press, 2013. ISBN:7030151232. (Chinese).

Li, D., Wang, S.L., Shi, W.Z., et al., On Spatial Data Mining and Knowledge Discovery (SDMKD). *Geomatics and Information Science of Wuhan University*, no. 6, 2001, pp. 22–30.

Li, R.H., and Belford, G.G., Instability of decision tree classification algorithms. *Proceedings of the 8th ACM SIGKDD International Conference on Knowledge Discovery and Data Mining*, Edmonton, Alberta, Canada, 2002, pp. 570–575.

Li, X.B., Sweigart, J., Teng, J., Donohue, J., and Thombs, L., A dynamic programming based pruning method for decision trees. *INFORMS Journal on Computing*, vol. 13, no. 4, 2001, pp. 332–344.

Lim, T.S., Loh, W.Y., and Shih, Y.S., An empirical comparison of decision trees and other classification methods. *Technical Report, No. 979. Department of Statistics, University of Wisconsin*, 1998.

Llorà, X., and Garrell, J.M., Evolution of decision trees. *Proceeding of the 4th Catalan Conference on Artificial Intelligence (CCIA'2001)*, ACIA Press, Washington, DC, September 26, 2001, pp. 115–122.

Lu, H., Setiono, R., and Liu, H., Effective data mining using neural networks. *IEEE Transactions on Knowledge and Data Engineering*, vol. 8, no. 6, 1996, pp. 957–961.

Lu, J.L., Wang, L.Z., Xiao, Q., et al., Incremental mining and evolutional analysis of co-locations. *Journal of Software*, vol. 25, no. 2, 2014, pp. 189–200.

Ma, L., Exploring the spatial data mining based on big data. *Journal of Xiang Yang Vocational and Technical College*, no. 6, 2017, pp. 80–82.

Mansour, Y., Pessimistic decision tree pruning based on tree size. In M. van Someren, and G. Widmer (Eds.), *Proceedings of the 9th European Conference on Machine Learning (ECML-97)*. Czech Republic, Berlin, Heidelberg: Springer Press, April 23–25, 1997, pp. 195–201.

May, M., and Savinov, A., An integrated platform for spatial data mining and interactive visual analysis. *Management Information Systems*, vol. 6, 2002, *Data Mining III*, pp. 51–60.

Mehta, M., Agrawal, R., and Rissanen, J., SLIQ: A fast scalable classifier for data mining. *Proceedings of the 5th International Conference on Extending Database Technology*, Avignon, France, March 25–29, 1996, pp. 18–32.

Mennis, J., and Guo, D.S., Spatial data mining and geographic knowledge discovery: An introduction. *Computers, Environment and Urban Systems*, vol. 33, no. 6, November 2009, pp. 403–408.

Mennis, J., and Liu, J.W., Mining association rules in spatio-temporal data: An analysis of urban socioeconomic and land cover change. *Transactions in GIS*, vol. 9, no. 1, 2010, pp. 5–17.

Miller, H., and Han, J., Geographic data mining and knowledge discovery: An overview. In H. Miller, and J. Han (Eds.), *Geographic data mining and knowledge discovery*. Boca Raton, FL: CRC Press, Taylor and Francis Group, 2001a, pp. 1–26.

Miller, H.J., and Han, J., Geographic data mining and knowledge discovery: An overview. In H.J. Miller, and J. Han (Eds.), *Geographic data mining and knowledge discovery*. London and New York: Taylor & Francis, 2001b, pp. 3–32.

Mingers, J., An empirical comparison of pruning methods for decision tree induction. *Machine Learning*, vol. 3, no. 4, 1989, pp. 319–342.

Mingers, J., Expert systems-rule induction with statistical data. *Journal of the Operational Research Society*, vol. 38, 1987, pp. 39–47.

Mugambi, E.M., Hunter, A., Oatley, G., and Kennedy, L., Polynomial-fuzzy decision tree structures for classifying medical data. *Knowledge-Based Systems*, vol. 17, no. 2–4, May 2004, pp. 81–87.

Ng, R.T., and Han, J.W., CLARANS: A method for clustering objects for spatial data mining. *IEEE Transactions on Knowledge and Data Engineering*, vol. 14, no. 5, September/October 2002, pp. 1003–1016.

Niblet, T., and Bratko, I., Learning decision rules in noisy domains. *Proceedings of Expert Systems*, vol. 86, 1986, pp. 25–34.

Olaru, C., and Wehenkel, L., A complete fuzzy decision tree technique. *Science Direct*, vol. 138, no. 2, September 2003, pp. 221–254.

Openshaw, S., Charlton, M., Wymer, C., and Craft, A., A mark 1 geographical analysis machine for the automated analysis of point data sets. *International Journal of Geographical Information Science*, vol. 1, no. 4, 1987, pp. 335–358.

Osei, B., and Kweku, M., Evaluation of decision trees: A multi-criteria approach. *Computers & Operations Research*, vol. 31, no. 11, September 2004, pp. 1933–1945.

Osei, B., and Kweku, M., Post-pruning in decision tree induction using multiple performance measures. *Computers & Operations Research*, vol. 34, no. 11, November 2007, pp. 3331–3345.

Papagelis, A., and Kalles, D., Breeding decision trees using evolutionary techniques. *Proceedings of the Eighteenth International Conference on Machine Learning (ICML 2001)*, Williams College, Williamstown, MA, USA, Morgan Kaufmann, June 28–July 1, 2001, pp. 393–400.

Quinlan, J.R., *C4.5: Programs for machine learning*. San Mateo, CA: Morgan Kaufmann, 1993, ISBN: 978-1-55860-238-0.

Quinlan, J.R., Induction of decision trees. *Machine Learning*, vol. 1, 1986, pp. 81–106.

Quinlan, J.R., Simplifying decision trees. *International Journal of Man-Machine Studies*, vol. 27, 1987, pp. 221–234.

Rastogi, R., and Shim, K. PUBLIC: A decision tree classifier that integrates building and pruning. *Proceedings of 24th International Conference on Very Large Data Bases*, New York, NY, USA, 2000, pp. 315–344.

Säuberlich, F., KDD und Data Mining als Hilfsmittel zur Entscheidungsunterstützung [KDD and data mining as aid for decision support]. *Dissertation Thesis*. Peter Lang, Frankfurt a. Main, Germany, 2000.

Shafer, J.C., Agrawal, R., and Mehta, M., SPRINT: A scalable parallel classifier for data mining. *Proceedings of 22nd International Conference on Very Large Data Bases*, Bombay, India, 1996, pp. 544–555.

Shekhar, S., Zhang, P., Huang, Y., and Vatsavai, R., Trends in spatial data mining. In H. Kargupta, A. Joshi, K. Sivakumar, and Y. Yesha (Eds.), *Data mining: Next generation challenges and future directions*. Cambridge, MA: AAAI/MIT Press, 2012, pp. 357–381.

Siegel, E.V., Competitively evolving decision trees against fixed training cases for natural language processing. In K. Kinnear (Ed.), *Advances in genetic programming*. Cambridge, MA: MIT Press, 1992. ISBN:9780262277181.

Tan, P.N., Michael, S., and Vipin, K., *Introduction to data mining*. Boston: Pearson Addison Wesley Press, 2006, ISBN 0-321-32136-7.

Turney, P.D., Cost-sensitive classification: Empirical evaluation of a hybrid genetic decision tree induction algorithm. *Journal of Artificial Intelligence Research*, vol. 2, 1994, pp. 369–409.

Verhein, F., and Al-Naymat, G., Fast mining of complex spatial co-location patterns using GLIMIT. *IEEE International Conference on Data Mining Workshops*, Omaha, NE, USA, 2007, pp. 679–684.

Wang, H.Z., and Peng, A.Q., Existing situation of data mining research and its development tendency. *Industry and Mine Automation*, vol. 37, no. 2, 2011, pp. 29–32.

Witten, I.H., and Frank, E., *Data mining-practical machine learning tools and techniques with Java implementation*. San Mateo, CA: Morgan Kaufmann, 2002. ISBN: 1558605525.

Wu, P.P., Wang, L.Z., and Zhou, Y.H., Discovering co-location from spatial data sets with fuzzy attributes. *Journal of Frontiers of Computer Science and Technology*, vol. 7, no. 4, 2013, pp. 348–358.

Yan, J., and Thill, J.C., Visual data mining in spatial interaction analysis with self-organizing maps. *Environment and Planning B*, vol. 36, 2009, pp. 466–486.

Yang, S.S., Wang, L.Z., and Lu, J.L., Primary exploration for mining spatial high utility co-location patterns. *Journal of Chinese Computer Systems*, vol. 35, no. 10, 2014, pp. 2302–2307.

Yao, X., and Thill, J.C., Neurofuzzy modeling of context-contingent proximity relations. *Geographical Analysis*, vol. 39, no. 2, 2007, pp. 169–194.

Yoo, J.S., and Shekhar, S., A joinless approach for mining spatial colocation patterns. *IEEE Transactions on Knowledge and Data Engineering*, vol. 18, no. 10, October 2006, pp. 1323–1337.

Yoo, J.S., and Shekhar, S., A partial join approach for mining co-location patterns: A summary of results, GIS'04. *Proceedings of the 12th Annual ACM International Workshop on Geographic Information Systems*, Washington, DC, USA, November 12–13, 2004, pp. 241–249. Copyright 2004 ACM 1-58113-979-9/04/0011.

Yoo, J.S., and Shekhar, S., A joinless approach for mining spatial colocation patterns. *IEEE Transactions on Knowledge and Data Engineering*, vol. 18, no. 10, 2006, pp. 1323–1337.

Yoo, J.S., Shekhar, S., and Mete, C., A join-less approach for co-location pattern mining: A summary of results. *Data Mining, Fifth IEEE International Conference on*, IEEE, Houston, TX, USA, November 27–30, 2005, pp. 813–816.

Yoo, J., Shekhar, S., Smith, J., and Kumquat, J., A partial join approach for mining collocation patterns. *Proceedings of the 12th Annual ACM International Workshop on Geographic Information Systems*, 2004, pp. 241–249.

Yu, W.H., Identifying and analyzing the prevalent regions of a co-location pattern using polygons clustering approach. *ISPRS International Journal of Geo Information*, vol. 6, no. 9, 2017, pp. 1–22.

Yuan, M., Use of knowledge acquisition to build wildfire representation in geographic information systems. *International Journal of Geographical Information Systems*, vol. 11, 1997, pp. 723–745.

Zhang, X., and Pazner, M., The icon image map technique for multivariate geospatial data visualization: Approach and software system. *Cartography and Geographic Information Science*, vol. 31, no. 1, 2004, pp. 29–41.

Zhou, G., Co-location decision tree for enhancing decision-making of pavement maintenance and rehabilitation. *Ph.D. Dissertation*, Virginia Tech, Blacksburg, Virginia, USA, 2011.

Zhou, G., Li, Q., and Deng, G., Maximal instance algorithm for fast mining of spatial co-location patterns. *Remote Sensing*, vol. 13, 2021, p. 960.

Zhou, G., and Wang, L., Integrating GIS and data mining to enhance the pavement management decision-making. *The 8th International Conference of Logistics and Transportation*, Chengdu, China, July 31–August 2, 2008.

Zhou, G., and Wang, L., GIS and data mining to enhance pavement rehabilitation decision-making. *Journal of Transportation Engineering*, vol. 136, no. 4, February 2010, pp. 332–341.

Zhou, G., and Wang, L., Co-location decision tree for enhancing decision-making of pavement maintenance and rehabilitation. *Transportation Research Part C*, vol. 21, no. 1, 2011, pp. 287–305.

Zhou, G., Zhang, R.T., Zhang, D.J., Norman, K., Soe, M., Clement, A., and Prasad, S.T., Manifold learning co-location decision tree for remotely sensed imagery classification. *Remote Sensing*, vol. 8, no. 10, 2016, p. 855.

2 Fundamentals of Mining Co-Location Patterns

2.1 BASIC CONCEPTS OF MINING CO-LOCATION PATTERNS

Spatial co-location pattern is a feature subset of the relationship between geographical instances. In spatial data, each spatial instance is recorded as $T.i$, where T is the spatial feature type of the spatial instance and i is the unique ID of the instance within each spatial feature type. $A.1$ in Figure 2.1, whose ID is 1, is an instance of spatial feature type A. The example of spatial data set in Figure 2.1 includes three types of spatial features A, B, and C. Among these instances, five instances belong to the spatial feature A, five instances belong to the spatial feature B, and four instances belong to the spatial feature C (Table 2.1).

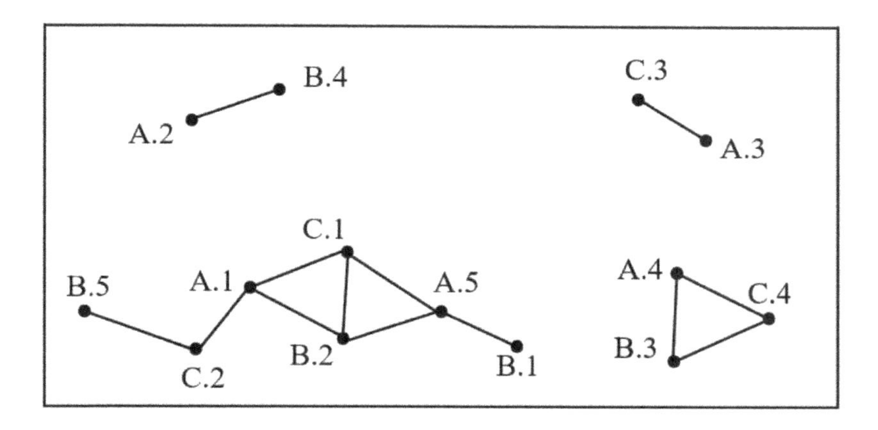

FIGURE 2.1 An example of spatial data set.

TABLE 2.1
Spatial Feature Types and Instances

Spatial feature types	Spatial instances
A	A.1, A.2, A.3, A.4, A.5
B	B.1, B.2, B.3, B.4, B.5
C	C.1, C.2, C.3, C.4

DOI: 10.1201/9781003139416-2

1. **Spatial neighbor relationships**

   A neighbor relationship is a Euclidean distance measure with a threshold of d, that is, $\mathrm{R}(A.1, B.1) \leftrightarrow (distance(A.1, B.1) \leq d)$ (Verhein and Al-Naymat 2007). In Figure 2.1, two instances of solid line connection meet the neighbor relationship definition. For example, $A.1$ and $B.2$ meet the neighbor relationship definition.

2. **Spatial co-location patterns**

   Spatial co-location patterns represent the subsets of features that are frequently located together in geographic space (Verhein and Al-Naymat 2007; Jiang et al. 2010; Hu and Qin 2008; Yu 2014). $\{A, B, C\}$ in Figure 2.1 is a co-location pattern.

3. **Clique**

   A clique is a group of objects such that all objects in that group are co-located with each other (Kim et al. 2011). For example, A.4, B.3, and C.4 form a clique in Figure 2.1.

4. **Neighborhood transaction**

   A neighborhood transaction (simply, transaction) is a set of instances T that forms a clique using a neighbor relationship R.

5. **Cut neighbor relation**

   A neighbor relation $r \in R$ between two event instances is called a cut neighbor relation if i_1 and i_2 are neighbors of each other but belong to distinct transactions.

6. **Row instance**

   A neighborhood instance I of a co-location C is a row instance (simply, instance) of C if I contains instances of all events in C and no proper subset of I does so.

7. **Table instance**

   The table instance of a co-location C is the collection of all row instances of C.

8. **IntraX row instance**

   A row instance I of a co-location C is an intraX row instance (simply, intraX instance) of C if all instances $i \in I$ belong to a common transaction T.

9. **IntraX table instance**

   The intraX table instance of C is the collection of all intraX row instances of C.

10. InterX row instance

A row instance I of a co-location C is an interX row instance (simply, interX instance) of C if all instances $i \in I$ have at least one cut neighbor relation. The interX table instance of C is the collection of all interX row instances of C.

11. InterX table instance

The interX table instance of C is the collection of all interX row instances of C.

12. Participation ratio

The participation ratio $\text{PR}(c, f_i)$ for feature type f_i in size-k co-location $\text{c} = \{f_1, \cdots, f_k\}$ is the fraction of instances of f_i which participate in any instance of co-location c (Verhein and Al-Naymat 2007; Jiang et al. 2010; Hu and Qin 2008; Yu 2014), i.e.

$$\text{PR}(c, f_i) = \frac{\pi_{f_i}\left(\left|table\_instance(c)\right|\right)}{\left|table\_instance(f_i)\right|}, \tag{2.1}$$

where π is a projection operation with duplication elimination.

13. Participation index

The participation index, Pi(C) is used as a co-location prevalence measure, i.e.

$$\text{PR}(c) = \min_{i=1}^{k}\{\text{PR}(c, f_i)\}. \tag{2.2}$$

14. Antimonotone

The participation ratio and participation index are antimonotone (monotonically nonincreasing) as the size of the co-location increases.

A negative co-location pattern refers to the patterns with strong negative correlation in spatial data and participation value less than the minimal frequency threshold (Jiang et al. 2010, Hu and Qin 2008). Jiang and others first proposed the concept of a negative co-location pattern and a mining algorithm (Jiang et al. 2010). The following is the concept of a negative co-location pattern.

15. Negative co-location pattern

A co-location pattern C is a negative co-location pattern, if $C = X \cup \overline{Y}$, where X is a set of positive items (positive spatial features), $\overline{Y}$ is a set of negative items (negative spatial features), and $\left|\overline{Y}\right| \geq 1, \quad X \cap Y = \varnothing$.

16. The PI of negative co-location patterns

The participation index of a negative co-location pattern $\text{C} = \text{X} \cup \overline{Y}$ is defined as $PI(C) = min_{i=1}^{k}\{PR(C, X_i)\}$, where $PR(C, X_i)$ is the participation ratio of spatial feature X_i in a negative co-location pattern C. The participation ratio $PR(C, X_i)$ can be calculated by:

$$PR(C, X_i) = \frac{\left|\pi_{X_i}\left(table\_instance(C)\right)\right|}{\left|table\_instance(X_i)\right|} \tag{2.3}$$

17. Prevalence negative co-location pattern

Given a minimum prevalence threshold (min_prev), a negative co-location pattern $C = X \cup \bar{Y}$ is a prevalence negative co-location pattern if C meets the following conditions: (a) $PI(X) \geq min\_prev$, $PI(Y) \geq min\_prev$ and $PI(X \cup Y) < min\_prev$; (b) $PI(C) \geq min\_prev$.

2.2 THREE BASIC TYPES OF CO-LOCATION PATTERN MINING ALGORITHMS

Huang, Shekhar, and Yoo proposed the join-based algorithm (Huang et al. 2004), the partial join algorithm (Hu and Qin 2008), and the joinless algorithm (Verhein and Al-Naymat 2007). These three basic algorithms not only solve the problem of spatial co-location pattern mining but also provide a basis for international scholars to deeply study co-location pattern mining. This section first introduces three basic types of co-location pattern mining algorithms in detail and analyzes their advantages and disadvantages, respectively.

2.2.1 Join-Based Algorithms

The join-based algorithm was first proposed by Huang, Shekhar, and Xiong (in Huang et al. 2004) for instances when the traditional transactional data mining method cannot be applied to spatial data sets. The join-based algorithm is also called the *transaction-free algorithm* because there is no self-defined transaction set in the spatial data set; this algorithm uses the principle of data mining to discover the co-location pattern in the spatial data on the basis of no transaction set. Literature (Yu 2014) introduces the concept of co-location patterns. As the basis of judging whether the co-location pattern is frequent, the participation index satisfies a non-monotonic property, that is, with the increase of the order of the co-location patterns, the participation index is monotonic and nonincreasing (Jiang et al. 2010; Huang et al. 2006). This property not only provides an effective pruning strategy to reduce unnecessary computing time but also ensures the correctness of co-location pattern mining.

The idea of the algorithm is: input spatial feature type set ET (Euclidean Type), spatial instance set E, user-defined proximity distance threshold and interest measurement threshold, and output frequent co-location pattern with a participation index greater than the user-defined interest measurement threshold. According to the definition of a participation index, it can be found that the participation index of all the size 1 co-location patterns is 1. This shows that all the size 1 co-location patterns are frequent, so it is not necessary to calculate their participation indexes or filter them based on frequency (Jiang et al. 2010). Therefore, both the size 1 candidate co-location pattern set and the frequent size 1 co-location pattern can be initialized to the spatial feature type set ET (Euclidean Type). The join-based algorithm has four basic steps: generating the candidate co-location pattern, generating the table instance of the candidate co-location pattern, pruning, and generating the co-location pattern.

Because the pruning process of the size 1 co-location patterns can be ignored, the iteration of the algorithm starts from the size 2 co-location patterns. The four steps of the algorithm are detailed as follows:

1. Generate candidate co-location pattern: frequent size k co-location patterns with $k - 1$ same instances are joined with each other to generate a size $k + 1$ candidate co-location pattern. This join strategy is the key of the algorithm. The apriori_gen function takes the size k frequent co-location pattern set P_k as the parameter.
2. Generate the table instance of the candidate co-location pattern: the generation of a table instance of the size $k + 1$ candidate co-location pattern also depends on the join operation. There are three strategies to calculate the total join, which are geometry strategy, combination strategy, and hybrid strategy. The geometric strategy can be realized by spatial join based on neighborhood relation, that is, the size k frequent co-location table instances are connected with the size 1 frequent co-location table instances. If in the previous step the size 2 frequent co-location pattern {A, B} and {A, C} connections generate the size 3 candidate co-location pattern {A, B, C}, then the table instance of {A, B} and the table instance of {A, C} can be connected to generate the table instance of {A, B, C}. This strategy is called a combination strategy. The hybrid strategy is to choose more effective geometry strategy and combination strategy in each iteration. Figure 2.2 details how to generate a row instance of co-location pattern {A, B, C} through join. Because the size k row instances are generated by joining the size $k - 1$ row instances. Although join-based algorithm can generate a complete and correct co-location pattern, the connection operations required in the generation process will increase with the increase of spatial feature types and their instances, and the calculation time will also increase.
3. Pruning: the candidate co-location pattern can be pruned by a given interest measurement threshold. First, the algorithm prunes co-locations based on the frequency, that is, only the candidate co-location pattern that is higher than the given interest measurement threshold is frequent. After the participation indexes of all candidate co-location patterns are calculated, pruning based on frequency is performed, and those non-frequent candidate co-location patterns will be deleted. Another pruning strategy is multi-resolution pruning. Multi-resolution pruning is learned from spatial data with coarse resolution using disjoint partitions. In the whole connection, the main pruning strategy

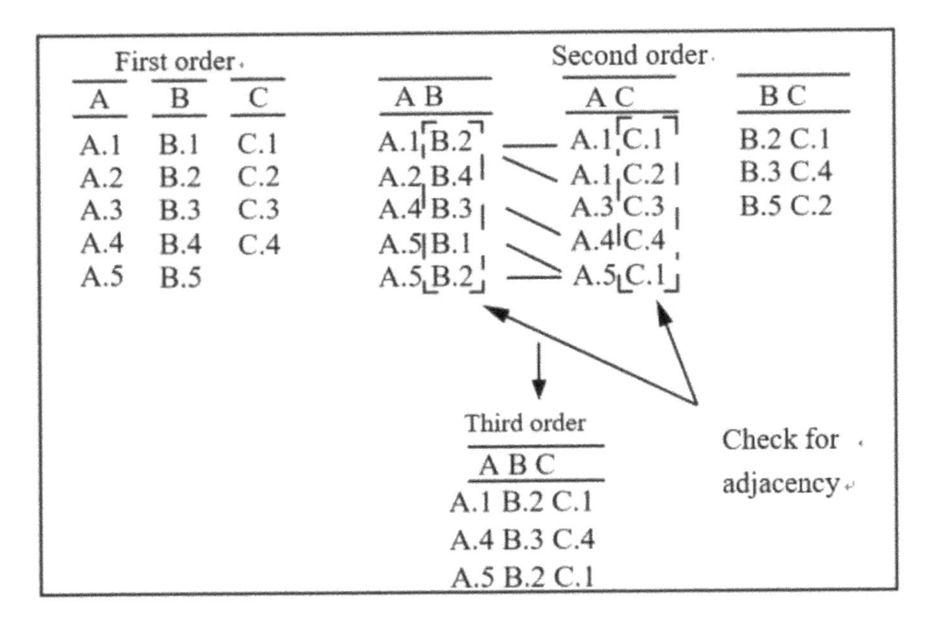

FIGURE 2.2 The example of generating a row instance of co-location patterns through join.

is based on frequency. This pruning strategy can ensure finding the complete and correct frequent co-location pattern.

4. Generate frequent co-location mode: through pruning, a co-location pattern satisfying a frequent threshold value greater than the user set value is selected.

The join-based algorithm is described as follows:

Input:

a. $E = \{Event - ID, Event - Type, Location\,in\,Space\}$ represents the set of instances of spatial features;
b. ET represents the set of spatial features;
c. R represents the neighbor relationship;
d. Θ represents the minimum frequency threshold, and α the minimum conditional probability threshold.

Output:

A set of co-locations with prevalence and conditional probability values greater than user-specified minimum prevalence and conditional probability threshold.

Variables

k : co-location size;

C_k : set of candidate size k co-locations;
T_k : set of table instances of co-locations in C_k;
P_k : set of prevalent size k co-locations;
R_k : set of co-location rules of size k;
T_C_k : set of coarse-level table instances of size k co-locations in C_k.

Steps

1. Co – location size k = 1, $C_1 = ET$, $P_1 = ET$;
2. $T_1 = generate\_table\_instance(C_1, E)$;
3. $if\ (fmul = TRUE)\ then$
4. $T\_C_1 = generate\_table\_instance(C_1, multi\_event(E))$;
5. Initialize data structure $C_k, T_k, P_k, R_k, T\_C_k$ to be empty for $1 < k \leq K$;
6. while$(not\ empty\ P_k\ and\ k < K)\ do\{$
7. $C_{k+1} = generate\_candidate\_colocation(C_k, k)$;
8. $if\ (fmul = TRUE)\ then$
9. $C_{k+1} = multi\_resolution\_pruning(\theta, C_{k+1}, T\_C_k, multi\_rel(R))$;
10. $T_{k+1} = generate\_table\_instance(\theta, C_{k+1}, T_k, R)$;
11. $P_{k+1} = select\_prevalent\_colocation(\theta, C_{k+1}, T_{k+1})$;
12. $R_{k+1} = generate\_cokocation\_rule(\alpha, P_{k+1}, T_{k+1})$;
13. k = k + 1;
14. }
15. return union$(R_2, \cdots, R_{k+1})$.

2.2.2 Partial Join Algorithms

A partial join algorithm, first proposed by Yoo and Shekhar (2004, 2006) and Yoo et al. (2005), is used to solve the problem of low performance caused by too many join operations in the join-based algorithm. The main idea of this algorithm is to optimize the algorithm by reducing the join operations and reduce the computing time. Before mining the co-location pattern, the partial join algorithm first transacts the spatial data set, that is, generating disjoint cliques in the data set. Generating a clique inevitably generates a segmentation neighborhood relationship (the "*cut neighbor relationship*" refers to the connection between two spatial instances if they belong to two different groups but meet the neighborhood relationship). The introduction of the cut neighbor relationship brings about new concepts: *intraX table instance*, *intraX row instance*, *interX table instance*, and *interX row instance*. In this way, each instance of a co-location pattern is divided into an interX row instance and an intraX row instance. A partial join algorithm is to join the interX and intraX instances and then calculate the participation of co-location pattern. The key of this algorithm is to reduce the number of neighbor relationships and divide them into as large groups as possible. Therefore, the method of the transactional spatial data set is very important, and the effect of the transaction is affected by the distribution of spatial instances. Since all spatial instances in a transaction are adjacent to each other, there is no need for spatial operations and composite operations; that is, the connection operations are not needed to find the candidate co-location pattern instances in the transaction. This method provides a framework for effective co-location pattern mining. The calculation cost of the instance join operation that only generates unidentified inter-group table instances in the transaction is relatively cheaper than that of the instance join operation that finds all co-location pattern table instances.

The basic steps of a partial join algorithm are as follows:

1. Transactionizing a spatial data set: given a spatial data set, the partial join algorithm first divides it to generate a transaction set. In reference (Hu and Qin 2008), some partition methods of generating a transaction set are introduced, such as the grid partition method, maximal clique method, minimal partition method, and so on. The ideal situation of a transactional spatial data set is to generate a group of the largest cliques while minimizing the number of sides divided by partitions. Figure 2.3 describes in detail the method of transactionizing a spatial data set in a partial join algorithm. The virtual coil represents the cluster, the diameter of each dotted circle is d (d is the distance threshold of neighbor relationship), the solid line represents that two instances meet the neighbor relationship, and the dotted line represents that two instances are a cut neighbor relationship.
2. Generate candidate co-location patterns: in the partial join algorithm, this step is the same as the join-based algorithm. It also uses the Apriori idea and join operation to generate size $k + 1$ candidate co-location pattern from the frequent size k co-location pattern.
3. Scan the transaction set to collect the intraX instances: scan the transaction set in each iteration and list the intraX instances of the candidate co-location

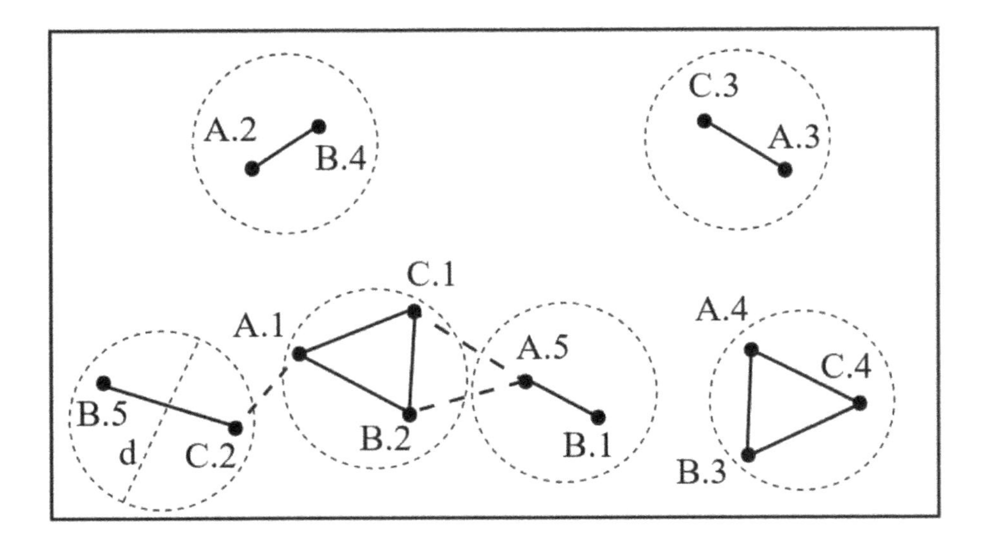

FIGURE 2.3 Transactionizing a spatial data set.

pattern. This step is similar to the Apriori algorithm. It should be noted that the transaction of spatial data set is different from that of market basket data. In traditional market basket data transactions, there are only Boolean item types, that is, items can only exist or not exist in transactions. In contrast, each item in a neighborhood transaction consists of a spatial feature type and its instances. A spatial feature type can have multiple instances in a transaction.

4. Generate interX table instances: similar to the method of generating table instance in join-based algorithm, the size k + 1 interX table instances are generated from the size k interX table instances.
5. Generate frequent co-location patterns: first merge the intraX table instance set and interX table instance set, then calculate the participation value of co-location pattern, then use the frequency threshold to prune and remove the non-frequent candidate co-location pattern.

Partial join algorithm is described as follows:

Input

a. E represents the set of spatial feature types;
b. $S = \{event\, instance\, id, event\, type, location\, in\, space\}$ represents the set of spatial instances;
c. R represents the neighbor relationship; and
d. Θ represents the minimum frequency threshold; α the minimum conditional probability threshold.

Output

A set of co-locations with prevalence and conditional probability values greater than the user-specified minimum prevalence and conditional probability threshold.

Variables

k : co-location size;
T : set of transactions
C_k : set of candidate size k co-locations;
P_k : set of prevalent size k co-locations;
R_k : set of co-location rules of size k;
$IntraX_k$: set of intraX table instances of co-locations in C_k;
$InterX_k$: set of interX table instances of co-locations in C_k and P_k;

Steps

1. $(T, InterX_2) = trasactionize(S, R)$;
2. $k = 1$; $C_1 = E$; $P_1 = E$;
3. $while(not\, empty\; P_k)do\{$
4. $C_{k+1} = generate\_candidate\_colocation(P_k)$;
5. $for\, all\, transaction\; t \in T$
6. $IntraX_{k+1} = gather\_intraX\_instances(C_{k+1}, t)$;

7. *if* $k \geq 2$
8. $InterX_{k+1} = gen\_interX\_instances(C_{k+1}, InterX_k, R)$;
9. $P_{k+1} = select\_prevalent\_colocation(C_{k+1}, IntraX_{k+1} \cup InterX_{k+1}, min\_prev)$;
10. $R_{k+1} = gen\_colocation\_rule(P_{k+1}, min\_cond\_prob)$;
11. $k = k+1$;
12. }
13. *return union* $(R_2, \cdots, R_{k+1})$.

2.2.3 Join-Less Algorithms

The joinless algorithm, first proposed by Yoo et al. (2005), is used to better solve the problem of excessive computing consumption caused by join operations. The main idea of the joinless algorithm is to establish a star neighborhood in a spatial data set. A star neighborhood is a circle centered on each spatial instance. The neighbor relationships among all spatial instances are stored in the star neighborhood, and the row instances of co-location patterns can be obtained by scanning the star neighborhood. The efficiency of joinless algorithms is higher than join-based algorithms and partial join algorithms because it uses the instance look-up method to get the instance of co-location patterns instead of the instance join method, which consumes too much calculation.

Figure 2.4 describes the star neighborhood partition method. The dashed circle in the figure is the star neighborhood of spatial instances *A*.1, *A*.4, and *C*.3 (that is, the star neighborhood with *A*.1, *A*.4, and *C*.3 as the center points). Instances within each star neighborhood are listed in Table 2.2. Although *A*.3 and *C*.3 satisfy the neighborhood relationship, there is no *A*.1 in the star neighborhood of *C*.3. This is because star neighborhood is a set of central instances and instances that satisfy the neighborhood relationship with the central instance and whose spatial feature types

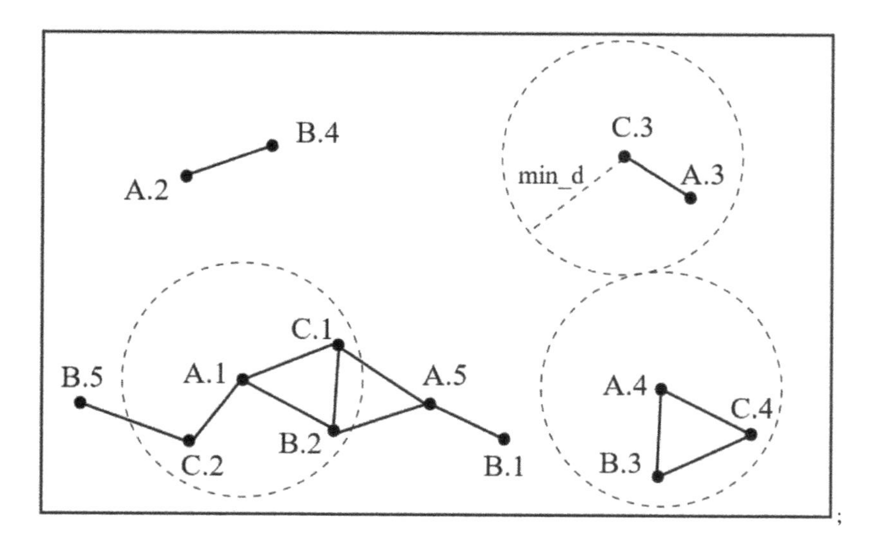

FIGURE 2.4 Star neighborhood partition method.

TABLE 2.2
Star Neighborhood

Central instance		Instances in star neighborhood
A	*A*.1	*A*.1, *B*.2, *C*.1, *C*.2
	A.2	*A*.2, *B*.4
	A.3	*A*.3, *C*.3
	A.4	*A*.4, *B*.3, *C*.4
	A.5	*A*.5, *B*.2, *B*.1, *C*.1
B	*B*.1	*B*.1
	B.2	*B*.2, *C*.1
	B.3	*B*.3, *C*.4
	B.4	*B*.4
	B.5	*B*.5, *C*.2
C	*C*.1	*C*.1
	C.2	*C*.2
	C.3	*C*.3
	C.4	*C*.4

are in order larger than those of the central instance. *A*.1 belongs to feature type A in order greater than *C*.3 belongs to feature type C.

The basic steps of joinless algorithm are as follows:

a. Transform the spatial data set into a non-intersecting star neighborhood set: given the input spatial data set and neighborhood relationship, first use the geometric method to find all adjacent object pairs. A star neighborhood can be generated from neighbor pairs by grouping neighbor objects of each object.
b. Generate candidate co-location patterns: first, all features are initialized to determine the size 1 prevalent co-location patterns according to the definition of the participation index. In the process of neighborhood materialization, the number of instances of each feature is known. Size k candidate co-location patterns are generated in size $k-1$ prevalent co-location patterns. These co-location patterns are filtered at the feature level. If any subset of the candidate co-location pattern is non-prevalent, the candidate pattern is pruned.
c. Filtering star instances of co-locations from star neighborhood sets: the star instances of the candidate co-locations are collected from the star neighborhood with the same feature type as the first feature of the co-locations. For example, instances of co-location pattern {B, C} are collected from the star neighborhood of feature B, and instances of co-location pattern {A, B, C} are collected from the star neighborhood of feature A. It should be noted that the number of candidate patterns examined in each star neighborhood is far less than the number of actual candidate patterns.

d. Choose the prevalent co-locations roughly: the star instances of size 2 are the cluster instance, because proximity is symmetric. For the co-locations above size 3, check whether a star instance is a cluster instance. Before this process, the co-locations are roughly filtered, that is, the participation indexes of candidate co-locations are calculated to filter them. If it is less than the minimum popularity threshold specified by the user, the candidate co-locations are pruned.
e. Filter row instances of co-locations: from star instances of candidate co-location patterns, use the instance lookup scheme to filter instances of co-locations.
f. Generate prevalent co-locations: the optimized filtering of co-locations is completed by the real participation index calculated from the instance of co-locations. Candidate co-locations that meet a given threshold are prevalent.

The joinless algorithm is described as follows:

Input

a. $F=\{f_1,\cdots,f_n\}$ represents the set of spatial features;
b. S represents the set of instances of spatial features;
c. R represents the neighbor relationship; and
d. Θ represents the minimum frequency threshold, and α the minimum conditional probability threshold.

Output :

A set of co-locations with prevalence and conditional probability values greater than user-specified minimum prevalence and conditional probability threshold.

Variables

$SN=\{SN_{f_1},\cdots,SN_{f_n}\}$: set of star neighborhood of spatial feature type f_i;
k : co-location size;
SI_k : set of star neighborhood of size k candidate co-locations;
CI_k : set of table instances of size k candidate co-locations;
P_k : set of prevalent size k co-locations; and
R_k : set of co-location rules of size k.

Steps

1. $SN = gen\_star\_neighborhoods(F,S,R)$;
2. $P_1 = F$; $k=2$;
3. $while(not\ empty\ P_{k-1})\ do$
4. $C_k = generate\_candidate\_colocations(P_{k-1})$;
5. $for\ i\ in\ 1\ to\ n\ do$
6. $for\ t \in SN_{f_i}\ where\ f_i = cf_1,\ cf_1\ is\ the\ first\ feature\ of\ C_k\ (cf_1,\cdots,cf_k)$.
7. $SI_k = filter\_star\_instances(C_k,t)$;
8. $end\ do$
9. $if\ k=2\ then\ CI_k = SI_k$
10. $else\ do\ C_k = select\_coarse\_prevalent\_colocations(C_k,SI_k,min\_prev)$
11. $CI_k = filter\_clique\_instances(C_k,SI_k)$;

12. *end do*
13. $P_k = select\_prevalent\_colocation(C_k, SI_k, min\_prev)$;
14. $R_k = gen\_colocation\_rule(P_k, min\_cond\_prob)$;
15. $k = k+1$;
16. *End do*
17. *return* $union(R_2, \cdots, R_k)$.

2.2.4 Advantages and Disadvantages of Three Basic Algorithms

The join-based algorithm uses the idea of data mining to find the neighbor relationships between spatial instances in continuous spatial data and then proposes a join strategy to generate instances of co-location patterns and candidate co-location patterns. In addition, the non-monotonic property of participation index is proposed and proved. One of the biggest advantages of the algorithm is that the join operations between the instances and the co-locations ensures that the full connection algorithm can mine the complete co-location pattern in the spatial data set, and the non-monotonic property of the participation index ensures that the join-based algorithm can mine the correct co-location patterns. However, the join operations in the join-based algorithm will increase with the increase of feature types and instances in the spatial data, and the amount of data in the spatial data set is often very large. Therefore, the join-based algorithm needs a lot of computing time to mine a complete and correct co-location pattern. In the era of big data, it is urgent to find useful knowledge quickly, so the time-consumption of the join-based algorithm cannot meet this requirement at the outset.

In order to solve the problem of excessive time consumption caused by too many join operations in the join-based algorithm, the partial join algorithm establishes disjoint clusters in the spatial data by transacting the spatial data set and divides the row instances of the spatial co-location patterns into interX row instances and intraX row instances. In this way, in order to generate co-locations, the partial join algorithm only needs to connect intraX instances and interX instances to calculate the participation index, which greatly reduces the number of join operations and the time consumption caused by the join. However, prior to mining the co-locations, the partial join algorithm needs to process the spatial data set, which also needs some extra time. Moreover, the transaction of spatial data sets is impacted by the distribution of spatial instances. If there are more cut neighbor relationships in this process, the efficiency of partial join algorithm will also be affected.

The advantage of the joinless algorithm is that it does not need the join between instances to generate co-locations. Similar to the partial join algorithm, the joinless algorithm needs to process the spatial data set before mining co-locations, that is, to establish the star neighborhood. By scanning the star neighborhood of each spatial instance, the neighborhood relationships between the instances can be obtained and all table instances can be generated. But the joinless algorithm is not perfect. It needs to scan the star neighborhood three times to generate table instances, which also consumes part of the computing time.

2.2.5 Other Algorithms

2.2.5.1 Co-Location Pattern Mining Algorithm with Rare Spatial Features

For co-location pattern mining, most of the research often emphasizes the equal participation of each spatial feature, which results in the inability to obtain interesting patterns involving spatial features with different frequencies. Therefore, Verhein and Al-Naymat (2007) studied the mining of co-location patterns with rare spatial features. First, the concept of maximum participation rate is put forward, and it is pointed out that compared with the participation rate, the maximum participation rate is also of great significance to co-locations; that is, it is also feasible to mine co-locations with rare spatial features by using the maximum participation rate. The maximum participation rate satisfies the weak monotonicity, which can effectively save computing time in the pruning stage. In Verhein et al. (2007), a large number of experiments are provided to evaluate the performance of two algorithms: Min-Max and maxPrune. Specifically, three cases are tested: (a) data sets that do not contain co-location patterns with rare spatial features, (b) data sets that contain patterns with rare spatial features, and (c) large data sets.

2.2.5.2 Maximal Clique Algorithm

Most of the algorithms of co-location pattern mining are based on the Apriori algorithm, which has two disadvantages. First of all, it is difficult to meaningfully include some types of complex relationships (especially negative ones) in a pattern. Second, the Apriori algorithm is slow. Kim et al. (2011) proposed the largest clique (not included in any other clique) to extract the complex maximal clique and then mine these largest cliques to obtain the interesting set of object types (including complex types). This means that the interesting complex relationships can be mined. At the same time, it is demonstrated that the application of GLIMIT (Geometrically Inspired Linear Itemset Mining in the Transpose) itemset mining algorithm to this task has better performance than the use of the Apriori style method.

In addition, the maximal clique algorithm is also studied by Wu et al. (2013), among others, and a polynomial algorithm called *AGSMC* (*Algorithm Generating Spatial Maximal Cliques*) is proposed to generate all the maximal cliques of general spatial data sets and to generate co-locations by using the maximal cliques. First, the existing advanced method is improved, which can extract all the neighborhood relationships between spatial objects. Second, a special tree data structure is proposed that can represent the maximal cliques. Third, a polynomial algorithm is proposed that can generate all the maximal cliques from general spatial data sets. It constructs a tree data structure and generates the maximal cliques by scanning the completed tree. The structure of the tree is as follows: for all nodes, only the filtered candidate nodes are generated as real nodes using advanced method. Fourth, through the experiment of a synthetic spatial data set and a real spatial data set, it is demonstrated that the algorithm with the polynomial is proper.

2.2.5.3 Density Based Co-Location Pattern Mining Algorithm

Most of the existing co-location pattern mining algorithms are generated and tested, that is, generating candidates and testing each candidate to determine whether it is a co-location pattern. In the test step, the candidate instances are identified to get their acceptability. Generally, the cost of case recognition is very high. In order to reduce the calculation of case recognition, a density based method is proposed in (Xiao et al. 2012). First, divide the object into multiple partitions, and then identify instances in dense partitions. The dynamic upper limit of a candidate's popularity remains unchanged. If the current upper limit is less than the threshold, it stops identifying its instances in the remaining partitions.

Figure 2.5 shows the partitioning method of the density based algorithm. The algorithm first divides the spatial data into grids, so that each grid contains different spatial instances, that is, the density of each grid is different. For high-density partition priority processing, it is possible to quickly find out the co-location mode and save a lot of time. For example, as shown in Figure 2.5, the sample data is divided into nine partitions, among which the density of partition 3, 5 and 7 is significantly higher than that of other partitions, so the spatial instances in these three partitions will be processed first.

2.2.5.4 Co-Location Pattern Mining Algorithm with Fuzzy Attributes

Fuzzy features refer to different types of things with some fuzzy attributes in space. In practical applications, spatial features not only contain spatial information but also attribute information, which is of great significance to decision making and knowledge discovery. If the attribute information is considered, the original co-location pattern mining algorithm is no longer used. Therefore, the concept of fuzzy feature and fuzzy co-location pattern is introduced in reference (Wu et al. 2013; He

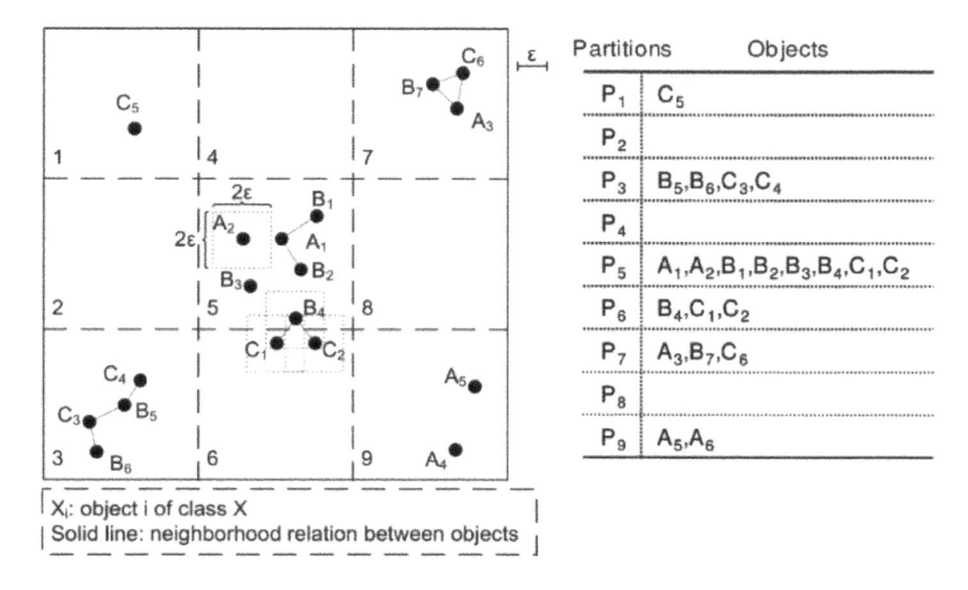

Partitions	Objects
P_1	C_5
P_2	
P_3	B_5, B_6, C_3, C_4
P_4	
P_5	$A_1, A_2, B_1, B_2, B_3, B_4, C_1, C_2$
P_6	B_4, C_1, C_2
P_7	A_3, B_7, C_6
P_8	
P_9	A_5, A_6

FIGURE 2.5 The method of spatial instance grid partition (Xiao et al. 2012).

2014), and a basic mining algorithm is proposed to mine co-location patterns with fuzzy attributes. Finally, the algorithm is evaluated by simulation data.

2.2.5.5 Co-Location Pattern Mining Algorithm with Time Constraints

The existing spatial co-location pattern mining algorithms often only use the location attribute of spatial data. But most spatial data has many different attributes, such as a time attribute and so on. There are two special cases in spatial data: time constraint and no user preset threshold. In order to solve the problem of co-location pattern mining in these two cases, literature (Zeng 2013) redefined the proximity between instances under time constraints, not only using Euclidean distance but also combining time constraints and Euclidean distance. In addition, the concept of no threshold is introduced into the data, which can mine co-location mode without a user preset threshold.

2.3 SPATIAL NEGATIVE CO-LOCATION MINING ALGORITHM

Spatial negative co-location pattern mining is of great significance in some applications, such as finding plant growth information. Botanists are not only interested in symbiotic information but also in mutually exclusive plant information. It can also be used to identify conflicting items (goods) in business data. However, only a few researchers have conducted the investigations on the mining of negative co-location patterns.

The reference (Jiang et al. 2010) was the first to propose an algorithm for mining negative co-location patterns. Jiang et al. (2010) and others put forward the definition of the participation index of negative co-location patterns. By analyzing the relationship between the participation index of the positive and negative co-locations, the calculation method of the participation index of the negative co-locations is proposed. The participation index of the negative co-locations provides a basis for judging whether a negative co-location pattern is prevalent. Although this algorithm successfully solves the problem of mining negative co-location patterns, it needs to mine all co-location patterns before mining negative co-location patterns (Li 2020). Therefore, the time consumption of the algorithm is too large to achieve fast and effective mining of the negative co-location mode.

The algorithm of negative co-location pattern mining proposed by Jiang et al. (2010) consists of the four steps:

a. obtaining the proximity between spatial instances;
b. generating size k candidate co-location patterns and negative co-location patterns;
c. generating size k prevalent co-location patterns; and
d. generating prevalent negative co-location patterns.

The generation of candidate negative co-locations is different from that of co-location pattern mining algorithms. The candidate negative co-location is generated by non-prevalent co-locations. In order to ensure that all co-locations and negative co-locations can be generated, the size k candidate co-locations in the algorithm is generated by joining size $k - 1$ prevalent co-locations and size $k - 1$ prevalent co-locations. The algorithm is described in this book.

2.4 DIFFERENCES BETWEEN POSITIVE AND NEGATIVE CO-LOCATION PATTERN MINING

Positive co-location patterns and negative co-location patterns are two different types of patterns in space. The knowledge contained in the two patterns plays a great role in people's understanding of spatial data and also has great application significances in many fields. First of all, from the definition point of view, a co-location pattern is a set of positively correlated instances in spatial data, that is, the instances contained in a co-location pattern are adjacent in space, while the negative co-location pattern is a set of instances with negative correlation in spatial data, and the instances in the pattern contain at least one negative correlation.

From the perspective of a participation index, the calculation method of a participation index of negative co-location patterns depends on the instances of participating in co-location patterns to calculate the participation index of negative co-location patterns. According to the definition of the participation index of negative co-location patterns, a method for calculating the participation index of negative co-location patterns is proposed by Jiang et al. (2010), that is, if $C = X \cup \overline{Y}$ is a candidate negative co-location, where $X = \{X_1, X_2, \cdots, X_k\}$. If $A_1, A_2, \cdots, A_k$ are the number of instances of spatial features $X_1, X_2, \cdots, X_k$ participating in co-location pattern $X \cup Y$, then the participation index of negative co-location pattern $X \cup \overline{Y}$ is:

$$PI(X \cup \overline{Y}) = min\left\{\frac{|X_1| - A_1}{|X_1|}, \frac{|X_2| - A_2}{|X_2|}, \cdots, \frac{|X_k| - A_k}{|X_k|}\right\}, \quad (2.4)$$

where $|X_i|$ is the number of instances of the spatial feature X_i.

From the perspective of mining algorithms, the ideas of the two algorithms are different. In co-location pattern mining, size k candidate co-locations are generated by connecting size $k - 1$ prevalent co-locations. In order to ensure that all co-location patterns and negative co-location patterns can be found, the size k candidate co-location patterns in the mining algorithm of negative co-location patterns are generated by the connection of size $k - 1$ prevalent co-location patterns and size 1 prevalent co-location patterns.

2.5 SUMMARY OF THIS CHAPTER

This chapter mainly introduces the basic theory of spatial co-location pattern mining. First, the concept of co-location patterns, row instances, the participation index, and other related concepts are given, which provides help for a better understanding of this book. Second, the related algorithms of spatial co-location pattern mining are reviewed. Spatial co-location pattern mining was a hot topic in recent years; there are many related research results, but the most classic co-location pattern mining algorithms are join-based, partial join algorithms, and joinless algorithms. Therefore, this chapter only focuses on these three algorithms.

REFERENCES

He, S., *The research on theory and application of buffer-based co-location spatial analysis.* Guilin: Guilin University of Technology, 2014.

Hu, C.P., and Qin, X.L., A novel positive and negative spatial co-location rules mining algorithm. *Journal of Chinese Computer Systems*, vol. 29, no. 1, 2008, pp. 80–84.

Huang, Y., Pei, J., and Xiong, H., Mining co-location patterns with rare events from spatial data sets. *GeoInformatica*, vol. 10, no. 3, 2006, pp. 239–260.

Huang, Y., Shekhar, S., and Xiong, H., Discovering colocation patterns from spatial data sets: A general approach. *IEEE Transactions on Knowledge and Data Engineering*, vol. 16, no. 12, 2004, pp. 1472–1485.

Jiang, Y., Wang, L., Lu, Y., et al., Discovering both positive and negative co-location rules from spatial data sets. *International Conference on Software Engineering & Data Mining*, IEEE, Chengdu China, June 23–25, 2010, pp. 398–403.

Kim, S.K., Kim, Y.H., and Kim, U., Maximal cliques generating algorithm for spatial colocation pattern mining. *FTRA International Conference on Secure & Trust Computing*, Loutraki, Greece, June 28–30, 2011, pp. 241–250.

Li, Q., Research on algorithm of positive/negative co-location pattern mining in spatial data. *MS. Thesis*, Guilin University of Technology, Guilin, China, July 2020.

Verhein, F., and Al-Naymat, G., Fast mining of complex spatial co-location patterns using GLIMIT. *IEEE International Conference on Data Mining Workshops*, IEEE, Omaha, NE, USA, 2007, pp. 679–684.

Wu, P.P., Wang, L.Z., and Zhou, Y.H., Discovering co-location from spatial data sets with fuzzy attributes. *Journal of Frontiers of Computer Science and Technology*, vol. 7, no. 4, 2013, pp. 348–358.

Xiao, X., Xie, X., and Ma, W.Y., Density-based co-location pattern discovery: DBLP, December 2012, pp. 1–10.

Yoo, J.S., and Shekhar, S., *A partial join approach for mining co-location patterns: A summary of results.* New York, USA: Association for Computing Machinery, GIS'04, November 12–13, 2004, Copyright 2004 ACM 1-58113-979-9/04/0011, pp. 241–249.

Yoo, J.S., and Shekhar, S., A joinless approach for mining spatial colocation patterns. *IEEE Transactions on Knowledge and Data Engineering*, vol. 18, no. 10, 2006, pp. 1323–1337.

Yoo, J.S., Shekhar, S., and Celik, M., A join-less approach for co-location pattern mining: A summary of results. *Data Mining, Fifth IEEE International Conference on*, IEEE, Houston, TX, USA, November 27–30, 2005, pp. 813–816.

Yu, C.L., Summary on algorithms for mining spatial co-location patterns. *Computer & Digital Engineering*, vol. 42, no. 7, 2014, pp. 1131–1136.

Zeng, X., *Research on co-location pattern mining without threshold under time constraint. Ph.D. Dissertation*, Yunnan University, 2013. (Chinese).

3 Principle of Mining Co-Location Patterns

3.1 INTRODUCTION

Chapter 1 introduced the fundamental theory of decision tree generation including tree construction, algorithm modeling, and attributes splitting criteria, pruning, and accuracy evaluation of decision tree performance. This chapter will present an innovative method called *co-location spatial decision tree induction*, which is to incorporate co-location (also called co-occurrence) mining into the decision tree. This chapter will describe the details of the co-location decision tree construction, algorithm, modeling, decision rule, node splitting criterion, node merging criterion, and leaf stopping criteria and then will give an example for illustrating the calculation process of the co-location spatial decision tree induction algorithm.

3.2 CO-LOCATION MINING ALGORITHMS

Huang et al. (2004) presented the first general framework of mining spatial co-location patterns. Afterward, Huang and her research team made further research and exploration into how mining co-location rules can be applied in spatial data analysis, spatial data pattern classification, and spatial geographic knowledge discovery. For example, Huang et al. (2005, 2006) adjusted the measure to treat cases with rare events, and Huang et al. (2008) used density ratios of different features to describe the neighborhood constraint together with a clustering approach. Xiao et al. (2008) presented a density based algorithm for mining a spatial co-location pattern, and Xiong et al. (2004) presented a buffer-based model to describe the neighborhood constraint for dealing with extended spatial objects such as lines and polygons.

On the other hand, different researchers have made efforts to improve the efficiency of the mining process of co-location. Yoo et al. (2004) proposed a partial-join algorithm. Yoo et al. (2005), Yoo and Shekhar (2006), and Wang et al. (2008) proposed a join-less algorithm and *N*-most prevalent collocated event in 2009 (Yoo and Bow 2009). Complex spatial co-location patterns are presented by Verhein and Al-Naymat (2007). Sheng et al. (2008) introduced the definition of an influence function based on the Gaussian kernel to describe the neighborhood constraint, in which the algorithm assumed a distribution of features on the global space. Hsiao et al. (2006) applied the spatial data mining of co-location patterns to support agriculture decision making. Zhang et al. (2004) enhanced the algorithm proposed in Huang et al. (2004) to mine special types of co-location relationships in addition to cliques, namely, the *spatial star* and *generic patterns*. Celik et al. (2006a) proposed the problem of mining mixed-drove spatial-temporal co-occurrence patterns (MDCOPs), which extends co-location pattern mining to the scope of

DOI: 10.1201/9781003139416-3

both time and space. Afterward, Celik et al. (2006b) further considered some constraints based on the result of MDCOP and the most top-k ranking issues, and Celik et al. (2007a, 2007b) partitioned a global space into small zones and applied the co-location mining algorithms on every zone for accumulated computation. Eick et al. (2008) also proposed to find regional co-location patterns based on clustering. Qian et al. (2009a) presented spatial co-location patterns with dynamic neighborhood constrain, and further spatial-temporal co-occurrence over zones (Qian et al. 2009b).

3.2.1 Definitions of the Co-Location Mining Method

The basic concept of spatial co-location, also called spatial co-occurrence, implies the presence of two or more spatial objects at the same location or at significantly close distances to each other. Co-location patterns can indicate interesting associations among spatial data objects with respect to their nonspatial attributes. In these methods, the neighborhood constraint is described by a distance threshold that is the maximal distance allowed for two events to be neighbors. Mathematically, co-location has been modeled by the references (Huang et al. 2004; Yoo and Shekhar 2006; Arunasalam et al. 2004; Zhou 2011; Zhou and Wang 2010, 2021):

Given

a. The training data is a set $S = \{s_1, s_2, \cdots, s_K\}$. Each sample $s_i = \{x_1, x_2, \cdots x_N\}$ is a vector, representing example-ID, spatial feature type, and location $\vee$, where location ∈ spatial framework. The training data is augmented with a vector $C = c_1, c_2, \ldots$ where $c_1, c_2, \ldots$ represent the class to which each sample belongs.
b. A neighbor relation $\Re$ over examples in S.

We have

a. A co-location, C, is defined as a subset of Boolean spatial features, $C \subseteq S$, whose instances form a clique under a neighbor relationship $\Re$. Usually, the neighbor relationship $\Re$ is a Euclidean distance metric. For example, if the two spatial objects satisfy the neighbor relationship, that is, distance $(s_i, s_j) \leq d$, they are called *neighbors*. If an instance shares co-location with another instance, the objects of all features form a clique relationship in the co-location.
b. Accompanying the co-location mining process, a co-location rule can be formed and expressed as $c_1 \Rightarrow c_2 \left(p, cp\right)$, where $c_1 \subseteq S$, $c_2 \subseteq T$, and $c_1 \cap c_2 = \Omega$; p is a number representing the prevalence measure, and cp is a number measuring conditional probability (Huang et al. 2004).

With the modeling given here, it can be noted that an important part in the co-location is proximity neighborhood, which is expressed using neighbor relation $\Re$. This relationship is based on the semantics of the application domains for forming a clique (Huang et al., 2006). For this reason, many researchers have presented different methods and algorithms to mode the neighbor relationship $\Re$, such as:

- Spatial relationships (e.g., connected, adjacent in GIS (Xiong et al. (2004));
- Metric relationships (e.g., Euclidean distance (Yoo and Shekhar 2006));
- Combined relationships (e.g., shortest-path distance in a graph such as a road-map); and
- Constrained relationships (e.g., Sheng et al. 2008,; Qian et al. 2009a, 2009b).

It is also noted that the R-proximity neighborhood concept is different from the neighborhood concept in topology, since some sets of an R-proximity neighborhood may not qualify to be R-proximity neighborhoods (Huang et al. 2006).

In order to describe the co-location algorithm, we first give several definitions (Huang et al., 2004).

A. *Participation ratio*

The participation ratio $pr(c, s_i)$ for feature type s_i in a size-k co-location $c = \{s_1, s_2, \cdots, s_K\}$ is the fraction of instances of feature s_i R-reachable to some row instance of co-location c -- $\{f_i\}$.

B. *Participation index*

The participation index pi (c) of a co-location c = $\{s_1, s_2, \cdots, s_K\}$ is $min_{i=1}^{k}\{pr(c_1, f_i)\}$. The participation index is used as the measure of prevalence of a co-location. The participation ratio can be computed as:

$$pi(c) = \frac{\pi_{s_i}(|table\_ins\tan ce(c)|)}{|table\_ins\tan ce(f_i)|} \tag{3.1}$$

Where π is the relational projection operation with duplication elimination.

C. *Conditional probability*

The conditional probability $cp(c_1 \Rightarrow c_2)$ of a co-location rule $c_1 \Rightarrow c_2$ is the fraction of row instances of c_1 $\Re$-reachable to some row instance of c_2. It is computed as:

$$cp = \frac{|\pi_{c1}(table\_instance(\{c_l \cup c_2\}))|}{|table\_instance(\{c_l\})|} \tag{3.2}$$

where π is the relational projection operation with duplication elimination.

3.2.2 Principle of Co-Location Pattern Mining Algorithms

Different types of co-location mining algorithms have been proposed in the past several years, for instance, He et al. (2008), Huang et al. (2003; 2004; 2005; 2006), Xiao et al.

(2008), Xiong et al. (2004), Yoo et al. (2005), Yoo and Shekhar (2006), Yoo and Bow (2009), Verhein and Al-Naymat (2007), Sheng et al. (2008), Celik et al. (2006a, 2006b, 2007a, 2007b), Qian et al. (2009a, 2009b). All of these proposed algorithms for mining co-location rules iteratively perform five basic tasks, namely (1) initialization, (2) determination of candidate co-locations, (3) determination of table instances of candidate co-locations, (4) pruning, and (5) generation of co-location rules. These tasks are carried out inside a loop iterating over the size of the co-locations.

3.2.2.1 Initialization

The task of initialization is to assign starting values to various data-structures. Obviously, the value of the participation index is 1 for all co-locations of size 1, that is, there is no need for either the computation of a prevalence measure or prevalence-based filtering, since all co-locations are prevalent.

3.2.2.2 Determination of Candidate Co-Locations

Determination of candidate co-location is usually realized using an approximate computation with rough threshold, so that a number of features with potential co-location can be found as much as possible. Huang et al. (2004) applied *apriori_gen*, proposed by Agarwal and Srikant (1994) to generate size $k + 1$ candidate co-locations from size k prevalent co-locations. This research will use only one geometric condition, spatial neighbor, to generate candidate co-location.

3.2.2.3 Determination of Table Instances of Candidate Co-Locations

The determination of table instances of candidate co-locations can be realized through a join query from $k + 1$ candidate co-location. The query takes the $k + 1$ candidate co-location set, C_{k+1}, and k prevalent co-locations in table instances as arguments and works.

In addition, during the join computation of generating table instances, Huang et al. (2004, 2006) presented three spatial neighbor relationship constraint conditions, the geometric approach (i.e., $(p.ins \tan ce_k, q.ins \tan ce_k) \in \Re$), a combinatorial distinct event-type constraint (i.e., $p.ins \tan ce_1 = q.ins \tan ce_1, \cdots, p.ins \tan ce_{k-1} = q.ins \tan ce_{k-1}$), and hybrid constraint, which combine the spatial neighbor relation constraint and combinatorial distinct event-type constraint. This research will adopt the hybrid constraint, but a slight modification will be made as follows:

- *Geometric constraint condition:* The geometric constraint condition will be neighborhood relationship-based spatial joins of table instances of prevalent co-locations of size k with table instance sets of prevalent co-locations of size 1. The spatial join operations consist of filter step and refinement. For these algorithms, Huang et al. (2004, 2006) has given a detailed description.
- *Event-type constraint condition:* The distinct event-type constraint is:

  Let $V = \{v_1, v_2, \cdots, v_c\}$ is a set of corresponding clusters center of feature a_1, $a_2, \ldots a_c$; the distinct event-type constrain is defined as:

$$\Gamma = \sum_{i=1}^{S}\sum_{k=1}^{c}\left(\left\|f_i - v_k\right\|\right)^2 \tag{3.3}$$

Where $\|x_i - v_k\|$ represents the Euclidean distance between f_i and v_k; Γ is a squared error clustering criterion. v_k, $\forall k = 1, 2, \cdots, c$ can be calculated by:

$$v_k = \sum_{i=1}^{N} f_i / N, \quad \mathrm{N} = 6; \forall k = 1, 2 \tag{3.4}$$

So, if the Γ is greater than a given threshold, Γ_θ, the i-th instance is assumed the distinct event.

3.2.2.4 Pruning

The purpose of pruning is to remove the non-prevalent co-locations from the candidate prevalent co-location set using the given threshold θ on the prevalence measure. Huang et al. (2004) proposed two basic pruning methods called *prevalence-based pruning method* and *multi-resolution pruning*. In this research, we will develop the spatial features pruning method. The multi-resolution pruning used the criterion of the coarse participation index based on the coarse table instance to eliminate the co-location. If its coarse participation indexes fall below the threshold, the co-location will be eliminated. This research will use the autocorrelation criterion of spatial features to eliminate the co-location features. The basic idea is:

> For a training data set $S = \{s_1, s_2, \cdots, s_K\}$, if the instance s_i and s_j are in co-location, where $s_i \in S$, $s_j \in S$ and $S_i = \{x_1, x_2, \cdots, x_N\}$, autocorrelation of the features vectors x_i and x_j, where $x_i \in S_i$ and $x_j \in S_j$, is calculated by:

$$\rho_{ij} = \frac{\sum_{i,j=1}^{N}\left(S_i - \bar{S}_i\right)\left(S_j - \bar{S}_j\right)}{\sqrt{\sum_{i,j=1}^{N}\left(S_i - \bar{S}_i\right)^2}\sqrt{\sum_{i,j=1}^{N}\left(S_j - \bar{S}_j\right)^2}} \tag{3.5}$$

If the two feature vectors are strongly cross-correlated when greater than a given threshold $T_{\rho_{ij}}$ in the training data, the co-location will be eliminated from the candidate co-location. Under this condition, a new neighbor relationship $\mathfrak{R}^p$ will have to be re-computed on the basis of the original relationship $\mathfrak{R}$ so that any two instances from each of the two partitions are $\mathfrak{R}$ neighbors. In this research, this computation is implemented under a local zone, that is, not a global extend.

3.2.2.5 Generating Co-Location Rules

Accompanying the generation of a co-location set, all the co-location rules with the user-defined conditional probability threshold from the prevalent co-locations and their table instances can be generated (Huang et al. 2004). The conditional probability of a co-location rule $cp(c_1 \Rightarrow c_2)$ in the event-centric model is the probability of c_1 reachable to a $\mathfrak{R}$-proximity neighborhood containing all the features in c_2.

An overview of the co-location mining algorithm is depicted in Figure 3.1.

```
Find-Co-Location Instance ()    /* function

Input:
a. Spatial data set
b. Criteria, including minimum prevalence threshold and other thresholds.

Output:
    A set of co-location rules

Variables Setup:
    k :  co-location size ¢
    C_k : set of candidate size-k co-locations
    T_k : set of table instances of co-location in C_k
    P_k : set of prevalent size
    R_k : set of co-location rules of size
    T_C_k : set of coarse-level table instances of size-k co-locations in C_k

Steps:
        Step 1: Co-location size-k =1;
        Step 2: IF (fmul=TRUE) THEN
                    T_C_1=generate_table_instance(C_1,multi_event);
        Step 3: While (not empty P_k and k < K ) do {
                    generate candidate co_location;
          IF (fmul=TRUE) THEN
              C_{k+1} = candidate size-k co-locations
          T_{k+1} = table instance of co-location in C_k
          P_{k+1} : = select prevalent colocation
          R_{k+1} : = generate co-location rule
          k = k + 1;
          }
    Step 4: return union
```

FIGURE 3.1 Overview of the co-location mining algorithm (modified from Huang et al. 2006).

3.3 CO-LOCATION DECISION TREE (CL-DT) ALGORITHMS

The basic idea of the presented co-location decision tree (CL-DT) algorithm is depicted in Figure 3.2. Co-location mining is used to induce the co-location rules. These induced co-location rules are used to guide the decision tree generation. The co-location mining algorithm was described in section 3.2. This section will focus on how a co-location decision tree is induced.

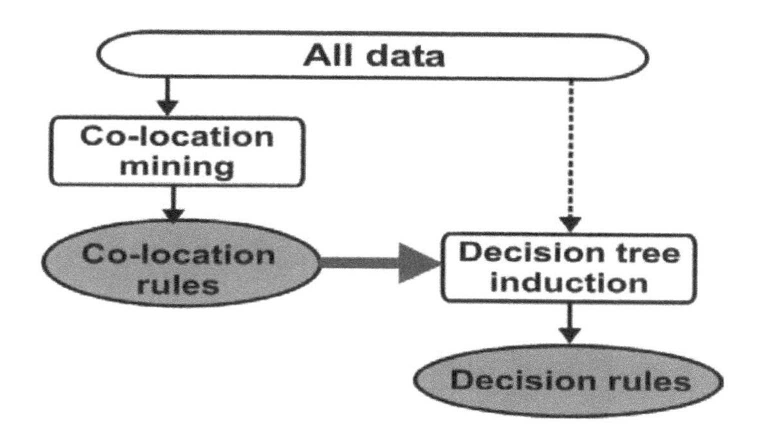

FIGURE 3.2 Flowchart of co-location decision tree induction.

3.3.1 CL-DT Algorithm Modeling

Let each sample $s_i = \{x_1, x_2, \cdots, x_N\}^T$ in data set $S = \{s_1, s_2, \cdots, s_d\}^T$ be a vector, representing example-ID, spatial feature type, and location Π, where d is the number of features, T is transpose, and the spatial location $\in$ spatial framework. The training data is augmented with a vector $C = \{c_1, c_2, \cdots, c_K\}$, where $\{c_1, c_2, \cdots, c_K\}$ represent the class to which each sample belongs. In order to assign an example to one of the classes, $C = \{c_1, c_2, \cdots, c_K\}$ $(K \geq 2)$, each internal node, m_i, carries out a decision or discriminant function, denoted by $g_{m_i}(x)$ for this purpose.

The functional of $g_{m_i}(x)$ varies due to various decision tree algorithms, such as univariate decision trees, linear multivariate decision trees, and nonlinear multivariate decision trees (Altincay 2007). This section discusses the generation of *univariate* co-location decision trees, and linear multivariate co-location will be discussed in section 3.4.

As usual, the CL-DT also utilizes a divide-and-conquer strategy to partition the instance space into decision regions by generating internal or test nodes. During the generation of the *univariate* decision trees, each internal node uses only one attribute to define a *decision* or a *model*. The mathematical model can be expressed by:

$$g_{m_i}(x) = s_i + b_{m_i} \tag{3.6}$$

Where b_{m_i} is a constant. The selection of the best attribute s_i, where $s_i \in S$, and the corresponding b_{m_i} for the instance subset reaching at the node b_{m_i} are the main tasks in the generation of the decision function.

As shown in Figure 3.3, the proposed algorithm for generating the CL-DT consists of a binary tree structure. At the beginning, the root node "accepts" all of examples, $S = \{s_1, s_2, \cdots, s_d\}^T$; the best feature is selected from input data set, and then splitting criterion is used for determining whether the root node will be split using a binary decision, **Yes** and **No**, with which the two intermediate nodes, noted by m_i, and m_j ($i = 1$ and $j = 2$ in Figure 3.3). For each of intermediate nodes, m_i and m_j, splitting

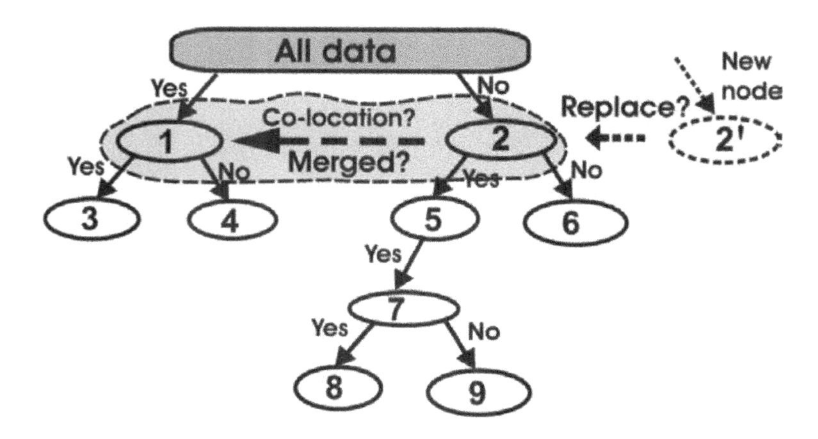

FIGURE 3.3 Co-location/co-occurrence decision tree.

criterion will be used to determine whether the node (e.g., m_i or m_j) should be further split. If **No**, this node is considered as a leaf node; then one of class labels is assigned to this leaf node. If **YES**, this node will be split by selecting one "best" feature. Once this "best" attribute is selected, the co-location criterion will be used to determine whether the sample with the "best" feature co-occurs with the sample with the previously selected features (see Figure 3.3). If **YES**, this node will be "merged" into the same classification as the co-location's, and one new "best" attribute will be selected again and re-determine whether the selected "best" feature co-occurs with the last best attribute; if **NO**, the node will further be split into a subset by repeating the previous work. This selection process is repeated until a non-co-occurrence feature is found.

This process continues recursively until all vectors are classified correctly. Finally, the termination criterion is satisfied; all leaf nodes are reached, and the class labels are assigned to each of the leaf nodes.

The outline of the algorithm is depicted in Figure 3.4. The input to this algorithm consists of the training records $S = \{s_1, s_2, \cdots, s_d\}^T$ and the attribute set $s_i = \{x_1, x_2, \cdots, x_N\}^T$. The algorithm works by recursively selecting the best attribute to split the data and expanding the intermediate nodes of the tree, and checking whether or not the attributes co-occur until the stopping criterion is met.

3.3.2 Attribute Selection

A pavement management database in fact contains many attributes, which are used to describe different pavement characteristics for various applications. This means that some of the attributes in the pavement management database do not in fact contribute to pavement rehabilitation-decision; that is, these attributes may be irrelevant to pavement decision making of maintenance and rehabilitation. Applications of these irrelevant attributes may cause negative influences to the pavement decision support or cause the decision tree to be over-fitted. Thus,

Input:
Training dataset D,
Splitting criterion,
Co-location threshold and criterion,
Terminal node threshold

Output:
A LC-DT decision tree with multiple condition attributes.

Process:
Step 1. Co-location mining
Step 2. Co-location rules
Step 3. Build an initial tree
Step 4. Starting with a single node, root. The root node includes all the rules and attributes.
Step 5. For each non-leaf node, e.g., m_i

- Perform label assignment test to determine if there are any labels that can be assigned.
- Take all the unused attributes in node m_i, and choose an attribute according to splitting criterion to further split m_i.
  - If the selected attribute satisfy the splitting criterion, partition the node into subsets.
  - If terminal condition is satisfied, stop splitting and assign m_i as a leaf node.

Step 6. For each of two non-leaf nodes in the same layer, e.g., m_i and m_j

- Apply co-occurrence algorithm, and test if the two nodes satisfy the co-occurrence criterion. If yes, merge two neighbor nodes; if no, please go head Step 5.

Step 7. Apply the algorithm recursively to each of the not-yet-stopped nodes, and update the *bottom nodes* in the tree built in Step 2.
Step 8. Generate decision rule by collecting decisions driven in individual nodes.
Step 9. The decision rules generated in Step 6 are used as initialization of co-location mining rule, and apply the algorithm of co-location mining rule to generate new associate rules.
Step 10. Re-organize the input data set, and repeat Step 2 through Step 7, until the classified results by the co-location mining rule and decision tree (rules) is consistent.

FIGURE 3.4 Outline of algorithm of co-location/co-occurrence decision tree (CL-DT).

to reduce the post-processing for obtaining an accurate and interpretable decision tree, these irrelevant attributes must be eliminated. Schetinin and Schult (2005) proposed *sequential feature selection* (*SFS*) algorithms based on a greedy heuristic to eliminate the irrelevant attributes. The basic idea of this method is a bottom-up search method, starting with one attribute and then iteratively adding the new attributes until a specified stopping criterion is met. The basic steps of the sequential feature selection are described in Figure 3.5.

Find_Best_Attribute () /* function
Step 1. Initiation with Set $i = 1$, $F_b = F_i = F_1$ /* W_b stands for the best feature
Step 2. Find the best attribute F_b
- Run the weighted linear tests $F_1, F_2, \ldots, F_T$ with the single attribute
- Select the test attribute F_k, $k \in T$
- Find the best test F_k, $k \in T$, **if** the test F_k is better than F_b, **then** $F_b = F_k$

Step 3. **if** the stopping criterion is met, **then** stop and return to F_b.
otherwise, $i := i + 1$, and **go to** Step 2.

FIGURE 3.5 The outline of steps of the SFS algorithm.

3.3.3 Co-Location Mining Rules

With the previously described co-location pattern mining operation, the co-location rules are traditionally generated with the user-defined conditional probability threshold from the prevalent co-locations and their table instances. The conditional probability of a co-location was given in section 3.2.1, that is

$$cp = \frac{\left|\pi_{c1}\left(table\_instance\left(\{c_l \cup c_2\}\right)\right)\right|}{\left|table\_instance\left(\{c_l\}\right)\right|} \tag{3.7}$$

Where π is the relational projection operation with duplication elimination.

However, this automatic method encountered problems, since conditional probability computation is time-consuming. Thus, this research manually forms the co-location rules by organizing individual decision making.

3.3.4 Node Merging Criteria

As mentioned earlier, in the pavement management database, some attributes are in co-occurrence in geography. For example, a set of co-occurrence attributes, *{car accident, traffic jam, police}* means when a car gets into an accident, the traffic jam will accompany occurrence and, further, police will arrive the accident site for cleaning up. So the three attributes co-occur frequently in a nearby region. If the three attributes are sequentially selected to generate the decision tree, the generated tree will be over-fitted. Thus, during the generation of the decision tree, the three nodes should be merged into one, or the other two nodes should be pruned.

One of the most major characteristics for the co-occurrence in a spatial database is that the attributes occur in nearby regions in geography for an event. For this reason, this research developed the following algorithm to "prune" the nodes.

For a spatial data set S, let $F = \{f_1, f_2, \cdots, f_k\}^T$ be a set of *spatial attributes*. Let $I = \{i_1, i_2, \cdots, i_n\}^T$ be a set of n instances in S, where each instance is a vector instance-ID, location, spatial features. The spatial attribute f_i, $f_i \subset F$ of instance i is denoted by $i.f$. We assume that the spatial attributes of an instance are from F and the location is within the spatial framework of the spatial database. Furthermore, we assume that there exists a neighbor relationship R in S. In addition, let, $V = \{v_1, v_2, \ldots, v_c\}$ is a set of corresponding clusters center in the data set S, where C is the number of clusters of spatial features, that is, $C \subseteq F$. To capture the concept of "nearby," the criterion of co-occurrence is defined as

$$\Pi_m = \sum_{i=1}^{C}\sum_{k=1}^{N} u_{ik}\left(\|x_i - v_k\|\right)^2$$

where $\|x_i - v_k\|$ represents the Euclidean distance between x_i and v_k; Π_m is a squared error-clustering criterion; $U = \{u_{ik}\}$, $i = 1, 2, \cdots, C$; $k = 1, 2, \cdots, C$. C is a matrix, and satisfies the following conditions:

$$u_{ik} \in [0,1], \quad \forall_i = 1,2\cdots,N, \quad \forall_j = 1,2\cdots,C \tag{3.8}$$

$$\sum_{k=1}^{C} u_{ik} = 1, \quad \forall_i = 1,2\cdots,N, \quad \forall_j = 1,2\cdots,C \tag{3.9}$$

So, if the Π_m is less than a given threshold, the two nodes are considered are in co-occurrence and thus should be merged.

3.3.5 Decision Rule Induction from CL-DT

After the co-location decision tree is generated, decision rules will be created by translating a decision tree into semantic expressions. Since a decision tree essentially partitions a data space into distinct disjoint regions via axis parallel surfaces created by its top-down sequence of decisions, decision rules will collect the individual decisions in each node through either top-down or down-up search.

Decision trees present a clear, logical model that can be understood easily by people who are not mathematically inclined (Wan et al. 2008).

3.4 LINEAR MULTIVARIATE CL-DT ALGORITHMS

The previous discussion is for *univariate* decision trees. In fact, the CL-DT algorithm can easily be extended to *linear multivariate* and/or multi-class trees. For a linear multivariate tree, the decision is based on a weighted linear combination of the features, which can be expressed by (Altincay 2007):

$$g_m(x) = \sum_{i=1}^{d} w_{mi}x_i + b_m \tag{3.10}$$

Similar to the univariate decision tree, the linear function at each node generates linear decision hyperplanes in the input space and separates the input space into two or multiple regions. For example, if a data set is partitioned into size-C classes, a maximum of C sub-nodes can be split, and up to $C(C-1)/2$ linear *multivariate* functions are constructed in each node. Correspondingly, $C(C-1)/2$ linear hyperplanes are constructed, thus separating each class from one another. It is also noted that an arbitrary hyperplane generated by a linear multivariate node is more powerful compared to the univariate case producing a hyperplane orthogonal to a particular axis (Altincay 2007). This process continues recursively until all vectors are classified correctly and a leaf node is reached.

3.5 EXAMPLE ANALYSIS

This section explains the details of a proposed algorithm and makes a comparison analysis between the proposed method and the C4.5 algorithm. Table 3.1 shows a data set of an extend example on the basis of the example adopted in Kervahut and Potvin (1996), where each instance is a member of class c_1, c_2 or c_3, and is described with four discrete attributes, namely a_1 with values f_{11}, f_{12}, f_{13}; a_2 with values f_{21}, f_{22}, f_{23}; a_3 with values $f_{31}, f_{32}, f_{33}, f_{35}, f_{36}$; and a_4 with values $f_{41}, f_{42}, f_{43}, f_{45}, f_{46}$ *(see Table 3.1).*

3.5.1 Decision Tree and Decision Rules Induction Using C4.5 Algorithms

The C4.5 algorithm builds decision trees from a data set of training data in the same way as ID3 (Agarwal and Srikant 1994; Al-Naymat 2008). At each node of the tree, C4.5 chooses one attribute of the data that most effectively splits its set of instances into subsets enriched in one class or the other. Its criterion is the normalized information gain (difference in entropy) that results from choosing an attribute for splitting the data. The attribute with the highest normalized information gain is chosen to make the decision. The C4.5 algorithm then repeats recursively the work on the smaller sub-lists. For the given example, the details are as follows:

TABLE 3.1
Data Set of Examples for Generating a Decision Tree and Co-Location Decision Tree

Example	Nonspatial attributes		Spatial attributes	Class results	
	a_1	a_2	a_3	C4.5 algorithm	Our algorithm
s_1	f_{11}	f_{21}	f_{31}	c_1	c_1
s_2	f_{11}	f_{22}	f_{32}	c_2	c_4
s_3	f_{12}	f_{22}	f_{33}	c_2	c_2
s_4	f_{12}	f_{23}	f_{32}	c_1	c_4
s_5	f_{13}	f_{21}	f_{35}	c_3	c_3
s_6	f_{13}	f_{22}	f_{36}	c_3	c_3

In this example, we have
$S = \{s_1, s_2, s_3, s_4, s_5, s_6\}$
$A = \{a_1, a_2, a_3\}$
$C = \{c_1, c_2, c_3\}$

Step 1. Start with the root as the current node to which the entire set of instance belongs.

Step 2. Select one attribute; evaluate the entropy of each subset of examples produced by splitting the set of examples at the current node along all possible attribute values. Then combine these entropy values into a global entropy value. For example, if we evaluate the entropy of attribute a_i, the set of examples S is partitioned into subsets $S_{i,j}$, Each subset $S_{i,j}$ contains the instances in S that share the same value $f_{i,j}$ for feature f_i. Then the entropy values of the subsets $S_{i,j}$ are combined to provide a single global value associated with attribute f_i, namely:

$$Gain(S, f_i) = E(S) - E(S, f_i) \quad \forall i = 1,2,3,4 \tag{3.11}$$

Where:

$$E(S, f_i) = -\sum_{f_{ij} \forall a_i} \left(\frac{|S_{ij}|}{|S|} \right) \times E(S_{ij}), \text{ and}$$

$$E(S) = \sum_{c_k \in C} \left(P_{S|c_k} \log_2 \left(P_{S|c_k} \right) \right)$$

Where:
S = the set of examples at the current node,
$|S|$ = the cardinality of set S,
C = the set of classes, and
$P_{S|c_k}$ = *the* proportion of examples in set S belonging to class c_k.

So, we have

$$\begin{aligned} E(S) &= E(S_{11}) + E(S_{12}) + E(S_{13}) \\ &= -0.5\log_2 0.5 - 0.5\log_2 0.5 - 0.0\log_2 0.0 - 0.5\log_2 0.5 \\ &\quad - 0.5\log_2 0.5 - 0.0\log_2 0.0 - 0.0\log_2 0.0 - 0.0\log_2 0.0 - 1.0\log_2 1.0 \\ &= 0.6934 \end{aligned}$$

$$\begin{aligned} E(a_1, S) &= \frac{2}{6}E(S_{11}) + \frac{2}{6}E(S_{12}) + \frac{2}{6}E(S_{13}) \\ &= \frac{2}{6}(-0.5\log_2 0.5 - 0.5\log_2 0.5 - 0.0\log_2 0.0) + \frac{2}{6}(-0.5\log_2 0.5 \\ &\quad - 0.5\log_2 0.5 - 0.0\log_2 0.0) + \frac{2}{6}(-0.0\log_2 0.0 - 0.0\log_2 0.0 - 1.0\log_2 1.0) \\ &= 0.4621 \end{aligned}$$

So, $Gain(S, x_i) = 0.6934–0.4621 = 0.2313$

Similarly, we can calculate the information gain for a2,

$$\begin{aligned} E(S,a_2) &= \frac{2}{6}E(S_{21})+\frac{3}{6}E(S_{22})+\frac{1}{6}E(S_{23}) \\ &= \frac{2}{6}(-0.5\log_2 0.5-0.0\log_2 0.0-1.0\log_2 1.0) \\ &\quad+\frac{3}{6}(-0.0\log_2 0.0-0.6667\log_2 0.6667-0.0\log_2 0.0) \\ &\quad+\frac{1}{6}(-0.5\log_2 0.5-0.333\log_2 0.333-0.0\log_2 0.0) \\ &= 0.5698 \end{aligned}$$

So, $Gain(S, x_i) = 0.98306 - 0.3698 = 0.6133$

$$\begin{aligned} E(S,a_3) &= \frac{2}{6}E(S_{21})+\frac{3}{6}E(S_{22})+\frac{1}{6}E(S_{23}) \\ &= 0.5698 \end{aligned}$$

$$\begin{aligned} E(S,a_4) &= \frac{2}{6}E(S_{21})+\frac{3}{6}E(S_{22})+\frac{1}{6}E(S_{23}) \\ &= 0.5698 \end{aligned}$$

Based on this computation of entropy, attribute a_1 is selected and the children of the root are created accordingly.

Step 3. Recursively apply this procedure to the children of the current node. The procedure stops at a given node when the node is homogeneous or when all attributes have been used along the path to this node. As shown in Figure 3.6, one child is homogeneous at $a_1 = f_{13}$, and no more processing is needed. The two other children are not homogeneous, and the procedure is recursively applied to each one of them, using the remaining attribute a_2.

Step 4. The stopping criterion is applied to check whether the procedure should be stopped. For this example, the final full decision tree can be created and is illustrated in Figure 3.6.

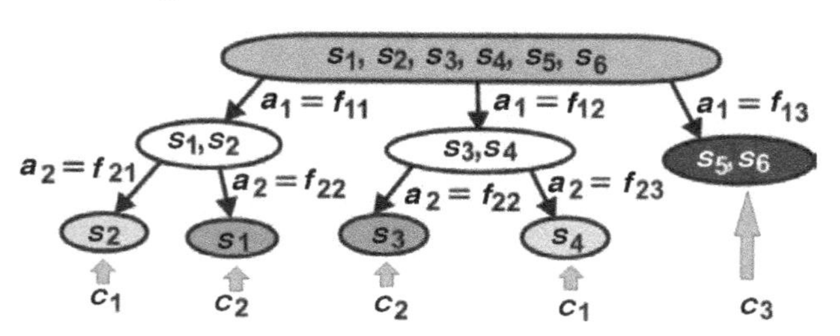

FIGURE 3.6 Decision tree induced by the C4.5 algorithm.

Step 5. With the generated decision tree in Figure 3.6, this decision tree encodes the following decision rules (see Figure 3.7):

IF $(f_1 = a_l)$ **THEN** c_3
IF $((f_1 = a_{12}$ and $f2 = a_{22})$ **OR** $(f_1 = a_{13}$ and $f_2 = a_{22}))$ **THEN** c_2
IF $((f_1 = a_{12}$ and $f2 = a_{23})$ **OR** $(f_1 = a_{13}$ and $f_2 = a_{21}))$ **THEN** c_1

FIGURE 3.7 Decision rules induced by C4.5 algorithm.

3.5.2 CL-DT Algorithms

Here, detailed steps for our algorithm would be presented. The proposed algorithm majorly includes two major steps: co-location mining rule induction and decision tree induction. The co-location mining rule induction mainly considers the spatial data and their characteristics, and decision tree induction mainly considers the non-spatial data. Integration of two data sets using two data mining technologies is in order to be complementary to the individual technologies' shortcoming. The steps of our algorithms are:

3.5.2.1 CL-DT Mining Rules

We first generate a co-location rule to discover which instances are "nearby", that is, having a neighborhood relationship. To this end, we follow up the steps described in section 3.2.2 as follows.

Step 1. Initialization: The purpose of initialization is to set up each variable and assign the memory size for each participation variable.

Step 2. Determination of candidate co-locations: The candidate instances with co-location relationships will be determined using the spatial neighborhood criterion with a given threshold, D_θ. In this particular example, the spatial neighborhoods for six instances is computed by:

$$Dist_{i,j} = \sqrt{\left(f_{3i} - f_{3j}\right)^2} \quad i, j \in 6 \tag{3.12}$$

With the given data set, the spatial distances of any one pair in this data set can form the following matrix:

$$Dist = \begin{Bmatrix} 0 & d_{12} & d_{13} & d_{14} & d_{15} \\ & 0 & d_{23} & d_{24} & d_{25} \\ & & 0 & d_{34} & d_{35} \\ & & & 0 & d_{45} \\ & & & & 0 \end{Bmatrix}$$

With the given values of instances S_2 and S_4, the matrix can be rewritten as follows:

$$Dist = \begin{bmatrix} 0 & d_{12} & d_{13} & d_{14} & d_{15} \\ & 0 & d_{23} & 0.0 & d_{25} \\ & & 0 & d_{34} & d_{35} \\ & & & 0 & d_{45} \\ & & & & 0 \end{bmatrix}$$

With this computation, instances S_2 and S_4 probably have co-location, since their spatial distance is equal to 0. Thus, S_2 and S_4 are listed as candidate co-locations.

Step 3. Determination of table instances of candidate co-locations: Based on the previously generated potential co-location instances, the determination of table instances of candidate co-locations will be implemented using a combination approach, that is, using spatial neighbor relationship constraint conditions (geometric approach) and distinct event-type constraints. The spatial geometric constrain is expressed thus:

$$d_{i,j} \leq D_\theta \quad d_{i,j} \subseteq Dist, \forall_i = 1, 2, \cdots, 6 \tag{3.13}$$

where D_θ is given threshold for spatial distance.

With the given example, the distinct event-type constraint is:

$$\Gamma = \sum_{i=1}^{6} \sum_{k=1}^{2} \left(\| f_i - v_k \| \right)^2 \tag{3.14}$$

where $\|x_i - v_k\|$ represents the Euclidean distance between f_i and v_k; $V = \{v_1, v_2\}$ is a set of corresponding clusters center of feature a_1 and a_2; Γ is a squared error clustering criterion. v_k, $\forall\, k = 1,2$ can be calculated by:

$$v_k = \sum_{i=1}^{N} f_i \,/\, N, \quad \text{N} = 6; \forall k = 1, 2 \tag{3.15}$$

So if the Γ is greater than a given threshold, Γ_θ, the i-th instance is assumed to be the distinct event.

Step 4. Pruning: As mentioned, this research used cross-correlation of the features vectors f_i and f_j to prune the candidate co-location. Since the features in this example have no correlation, the pruning is unnecessary. si s_j

Step 5. Generating Co-location Rules: Based on the described co-location mining approach, the co-location rules from the prevalent co-locations and their table instances can be generated. They are depicted in Figure 3.8.

IF ($d_{i,j} \le D_\theta$) **THEN** s_i and s_j are potentially co-located
IF ($d_{i,j} \le D_\theta$) and $\Gamma \le T_\theta$) **THEN** s_i and s_j co-location
IF (s_i and s_j are co-located) **THEN** c_2
OTHERWISE, c_1

FIGURE 3.8 Co-location mining rule.

3.5.2.2 CL-DT Induction

With the previously described co-location mining rule, decision tree induction will be carried out on the basis of the induced co-location mining. Thus, during the generation of the decision tree at this time, the co-location mining rule will constrain the process of the decision tree induction. The steps are as follows:

Step 1. Start with the root as the current node to which the entire set of instance belongs.

Step 2. With the similar computation of the entropy of each subset of instances produced by splitting the set of instances at the root node, attribute a_1 is selected.

Step 3. With the selected attribute, a_1, split the instances along the path to this node. As noted, one child is homogeneous at $a_1 = f_{13}$, which implies that no further processing is needed. The two other children are not homogeneous, and the procedure is recursively applied to each one of them, using the remaining attribute a_2.

Step 4. During the recursive procedures to attribute a_2, the process will automatically recall the co-location mining rule; that is, instances, s_2 and s_4, are co-located, i.e., co-occurred. Thus the s_2 and s_4 must be the same class.

Step 5. The stopping criterion is applied to check whether the procedure should be stopped. For this example, the final full decision tree can be created and is illustrated in Figure 3.9.

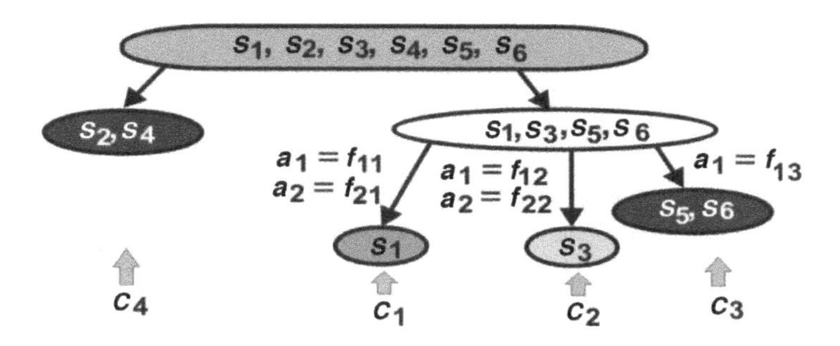

FIGURE 3.9 Decision tree induced by our algorithm.

Step 6. With the generated decision tree described, this decision tree encodes the following decision rules (Figure 3.10):

IF $((s_2, s_4) = \text{co-location})$ **THEN** c_4
IF $(a_1 = f_{13})$ **THEN** c_3
IF $(a_1 = f_{11}$ and $a_2 = f_{21})$ **THEN** c_1
IF $((a_1 = f_{12})$ and $a_2 = f_{22})$ **THEN** c_2

FIGURE 3.10 Decision rules induced by our algorithm.

3.6 DISCUSSION AND ANALYSIS OF CL-DT

As observed from the previous two examples, the C4.5 algorithm is very sensitive to the entropy formula. If we selected attribute f_2 before f_1, a different tree may be created in the example. Therefore, it can be imagined that many different decision trees may be generated when modifying the entropy formula (Tan et al. 2006). On the other hand, one major weakness of the C4.5 algorithm is that a node is created for each value of a given attribute. As mentioned before, a few attributes are co-occurrent of one another, that is, only a single attribute can get a good global evaluation in some cases, even if its entropy is good only for a few values among all its possible values (Kervahut and Potvin 1996), where the entropy of an attribute is computed as a linear weighted sum over all values.

The CL-DT uses a co-location mining technology to first classify the co-location attributes. This is in fact equivalent to pruning the nodes whose attributes co-occur with the previous attributes. Consequently, this proposed CL-DT overcomes the weakness of the C4.5 algorithm, which creates a node for each value of a given attribute. Obviously, the proposed CL-DT has the capability of handling rare events, which may arise naturally in the original data set because of the lower probability of occurrence of certain classes or the shortage of data for certain classes. Obviously, the CL-DT inherits all the advantages from regular decision trees, such as the recursive divide-and-conquer approach and efficient tree structure for rule extraction. Moreover, the proposed CL-DT allows it to solve classification problems with co-location and co-occurrence classes, making it more robust in real-world situations.

The quality of a decision tree is based on both its accuracy and complexity. The accuracy is assessed by testing the induced decision tree and/or decision rules with a new data set and then comparing the predicted classes with the real classes. The complexity is related to the shape and size of the tree. Obviously, the proposed CL-DT is capable of creating a simple and highly accurate decision tree because this algorithm has used the co-occurrence mining rule as the initialization to induce the decision tree and decision rule. However, most classification algorithms sought the models that attained the highest accuracy or, equivalently, the lowest error rate but with a complex tree and rules. For the same accuracy, simple trees are preferred over complex ones.

Traditionally, most of the decision tree induction algorithms have not been capable of producing compact solutions, that is, free expansion during generation of a decision tree, despite the adoption of pruning. On the other hand, since the decision tree is freely expanded, the decision rules also are freely expanded because the decision rules directly capture the individual decision of each node. These rules essentially correspond to decision regions that overlap each other in the data space. The proposed CL-DT is capable of create a compact solutions for decision trees and decision rules.

REFERENCES

Agarwal, R., and Srikant, R., Fast algorithms for mining association rules. *Proc. of the 20th Int'l Conference on Very Large Data Bases*, Santiago, Chile, September 12, 1994, pp. 487–499.

Al-Naymat, G., Enumeration of maximal clique for mining spatial co-location patterns. *AICCSA 08–6th IEEE/ACS International Conference on Computer Systems and Applications*, Doha, Qatar, 2008, pp. 126–133.

Altincay, H., Decision trees using model ensemble-based nodes. *Pattern Recognition*, vol. 40, December 2007, pp. 3540–3551.

Arunasalam, B., Sanjay, C., Sun, P., and Munro, R., Mining complex relationships in the SDSS SkyServer spatial database. *Proceedings in International Computer Software and Applications Conference*, Hong Kong, China, vol. 2, 2004, pp. 142–145.

Celik, M., Shekhar, S., Rogers, J., Shine, J., and Yoo, J., Mixed-drove spatio-temporal co-occurrence pattern mining: A summary of results. *Proceedings of the 6th International Conference on Data Mining*, Hong Kong, China, 2006a, pp. 119–128.

Celik, M., Shekhar, S., Rogers, J., and Shine, J., Sustained emerging spatio-temporal co-occurrence pattern mining: A summary of results. *Proceedings of the 18th IEEE International Conference on Tools with Artificial Intelligence*, Arlington, VA, USA, 2006b, pp. 106–115.

Celik, M., Shekhar, S., Rogers, J., Shine, J., and Kang, J., Mining at most top-K mixed-drove spatio-temporal co-occurrence patterns: A summary of results. *Proceedings of the 23rd IEEE International Conference on Data Engineering Workshop*, Istanbul, Turkey, 2007a, pp. 565–574.

Celik, M., Kang, J., and Shekhar, S., Zonal co-location pattern discovery with dynamic parameters. *Proceedings of the 7th IEEE International Conference on Data Mining*, Omaha, NE, USA, 2007b, pp. 433–438.

Eick, C.F., Ding, W., Stepinski, T.F., Nicot, J.P., and Parmar, R., Finding regional co-location patterns for sets of continuous variables in spatial datasets. *Proceedings of the ACM International Symposium on Advances in Geographic Information Systems*, Irvine, CA, New York, 2008, pp. 260–269.

He, J.F., He, Q.M., Qian, F., and Chen, Q., Incremental maintenance of discovered spatial colocation patterns. *Proceedings-IEEE International Conference on Data Mining Workshops, ICDM Workshops*, Pisa, Italy, 2008, pp. 399–407.

Hsiao, H.W., Tsai, M.S., and Wang, S.C., Spatial data mining of colocation patterns for decision support in agriculture. *Asian Journal of Health and Information Sciences*, vol. 1, no. 1, 2006, pp. 61–72.

Huang, Y., Pei, J., and Xiong, H., Mining co-location patterns with rare events from spatial data sets. *GeoInformatica*, vol. 10, no. 3, September 2006, pp. 239–260.

Huang, Y., Shekhar, S., and Xiong, H., Discovering colocation patterns from spatial data sets: A general approach. *IEEE Transactions on Knowledge and Data Engineering*, vol. 16, no. 12, December 2004, pp. 1472–1485.

Huang, Y., Xiong, H., Shekhar, S., and Pei, J., Mining confident co-location rules without a support threshold. *Proceedings of the ACM Symposium on Applied Computing*, New York, NY, USA, 2003, pp. 497–501.

Huang, Y., Zhang, L.Q., and Yu, P., Can we apply projection Based frequent pattern mining paradigm to spatial co-location mining. *Lecture Notes in Computer Science*, vol. 3518, LNAI, 2005, pp. 719–725.

Huang, Y., Zhang, P.S., and Zhang, C.Y., On the relationships between clustering and spatial co-location pattern mining. *International Journal on Artificial Intelligence Tools*, vol. 17, no. 1, Arlington, VA, USA, February 2008, pp. 55–70.

Kervahut, T., and Potvin, J.Y., An interactive-graphic environment for automatic generation of decision trees. *Decision Support Systems*, vol. 18, 1996, pp. 117–134.

Qian, F., He, Q.M., and He, J.F., Mining spatial co-location patterns with dynamic neighborhood constraint. *Lecture Notes in Computer Science (Including Subseries Lecture Notes in Artificial Intelligence and Lecture Notes in Bioinformatics, Bled, Slovenia)*, vol. 5782, LNAI, no. PART 2, 2009a, pp. 238–253.

Qian, F., He, Q.M., and He, J.F., Mining spread patterns of spatio-temporal co-occurrences over zones. *Lecture Notes in Computer Science (Including Subseries Lecture Notes in Artificial Intelligence and Lecture Notes in Bioinformatics, Seoul, Korea)*, vol. 5593, no. PART 2, 2009b, pp. 677–692.

Schetinin, V., and Schult, J., A neural-network technique to learn concepts from electroencephalograms. *Theory in Biosciences*, vol. 124, no. 1, August 15, 2005, pp. 41–53.

Sheng, C., Hsu, W., Lee, L., and Tung, M.A., Discovering spatial interaction patterns. In J.R. Haritsa, R. Kotagiri, and V. Pudi (Eds.), *DASFAA 2008: LNCS*, vol. 4947. Heidelberg: Springer, 2008, pp. 95–109.

Tan, P.N., Michael, S., and Vipin, K. *Introduction to data mining*. Boston, MA: Pearson Addison Wesley Press, 2006, ISBN 0-321-32136-7.

Verhein, F., and Al-Naymat, G., Fast mining of complex spatial co-location patterns using GLIMIT. *Proceedings of IEEE International Conference on Data Mining, ICDM*, Omaha, NE, USA, 2007, pp. 679–684.

Wan, Y., Zhou, J.G., and Bian, F.L., CODEM: A novel spatial co-location and de-location patterns mining algorithm. *Proceedings in 5th International Conference on Fuzzy Systems and Knowledge Discovery*, Ji'nan, China, *FSKD 2008*, vol. 2, 2008, pp. 576–580.

Wang, L.Z., Bao, Y.Z., Lu, J., and Yip, J., A new join-less approach for co-location pattern mining. *Proceedings of IEEE 8th International Conference on Computer and Information Technology*, CIT, Sydney, NSW, Australia, 2008, pp. 197–202.

Xiao, X., Xie, X., Luo, Q., and Ma, W., Density based co-location pattern discovery. *Proceedings of the 16th ACM SIGSPATIAL International Conference on Advances in Geographic Information Systems*, Irvine, CA, USA, November 5, 2008, pp. 1–10.

Xiong, H., Shekhar, S., Huang, Y., Kumar, V., Ma, X.B., and Yoo, J.S., A framework for discovering co-location patterns in data sets with extended spatial objects. *Proceedings of the Fourth SIAM International Conference on Data Mining*, Visto, FL, 2004, pp. 78–89.

Yoo, J.S., and Bow, M., Finding N-most prevalent colocated event sets: Source: *Lecture Notes in Computer Science*, vol. 5691 LNCS. *Data Warehousing and Knowledge Discovery: 11th International Conference, DaWaK 2009, Proceedings*, Linz, Austria, 2009, pp. 415–427.

Yoo, J.S., and Shekhar, S., A joinless approach for mining spatial colocation patterns. *IEEE Transactions on Knowledge and Data Engineering*, vol. 18, 2006, no. 10, October, pp. 1323–1337.

Yoo, J.S., Shekhar, S., and Celik, M., A join-less approach for co-location pattern mining: A summary of results. *Proceedings of IEEE International Conference on Data Mining, ICDM*, Houston, TX, USA, 2005, pp. 813–816.

Yoo, J., Shekhar, S., Smith, J., and Kumquat, J., A partial join approach for mining collocation patterns. *Proceedings of the 12th Annual ACM International Workshop on Geographic Information Systems*, New York, NY, USA, 2004, pp. 241–249.

Zhang, X.N., Mamoulis, D., Cheung, W., and Shou, Y., Fast mining of spatial collocations. *Proceedings of the Tenth ACM SIGKDD International Conference on Knowledge Discovery and Data Mining*, ACM Press, New York, 2004, pp. 384–393.

Zhou, G., Co-location decision tree for enhancing decision-making of pavement maintenance and rehabilitation. *Ph.D. Dissertation*, Virginia Tech, Blacksburg, Virginia, USA, 2011.

Zhou, G., and Wang, L., GIS and data mining to enhance pavement rehabilitation decision-making. *Journal of Transportation Engineering*, vol. 136, no. 4, February 2010, pp. 332–341.

Zhou, G., and Wang, L., Co-location decision tree for enhancing decision-making of pavement maintenance and rehabilitation. *Transportation Research Part C*, vol. 21, no. 1, April 2012, pp. 287–305.

4 Manifold Learning Co-Location Pattern Mining

4.1 INTRODUCTION

In the past a few decades, the decision tree (DT) induction method has become one of the most prevalent and powerful techniques for data mining (Appel et al. 2013; Witten and Frank 2000; Simard et al. 2000; Farid et al. 2014; Franklin et al. 2001; Franco-Arcega et al. 2011; Greiner and Hormann 1998; Huang et al. 2009; Lacar et al. 2001; Mansour 1997; Mohammad et al. 2002; Moustakidis et al. 2012; Osei-Bryson 2007, 2008; Xu and Anwar 2013; Chasmer et al. 2014; Polat and Gunes 2009; Zhang 2015; Zhang et al. 2015). As for the defects of traditional DTs, Zhou (2011) and Zhou and Wang (2012) presented a co-location-based decision tree (CL-DT) method to enhance decision making and has been successfully applied in pavement maintenance strategies. The major characteristics of CL-DT consider the geospatial relationship of these attribute data in addition to the traditional attribute data and use a co-location mining technology to first classify the co-location attributes. The basic idea of CL-DT is to apply co-location rules to induce the generation of DT. The processes of CL-DT consist of (a) selecting nonspatial and spatial data, (b) determining rough candidate co-locations, (c) determining candidate co-locations, (d) pruning the non-prevalent co-locations, (e) inducing co-location rules, (f) formulating node merging criterion, and (g) inducing a co-location decision tree.

However, the CL-DT uses Euclidean distance as a geometric constraint condition to refine the candidate co-location instances; that is,

$$dist_{i,j} = \sqrt{\left(X_i - X_j\right)^2 + \left(Y_i - Y_j\right)^2} \quad \forall \mathrm{i}, \mathrm{j} = 1, 2, \cdots, \mathrm{N} \tag{4.1}$$

where X, Y are spatial data in the database. In fact, although Euclidean distance can effectively represent the real distance between instances when they belong to a linear distribution in three-dimensional (3D) space, it is not capable of representing the real distance when they have a nonlinear distribution in 3D or higher-dimensional space (Zhan and Hua 2005). For example, Figure 4.1a shows a data set of a Swiss roll in a 3D space, in which the Euclidean distance $\overline{AB}$ between A and B is shorter than the real distance $\widehat{ACB}$, which is along the surface of the roll. If the Swiss roll is unfolded into Figure 4.1b using the *maximum variance unfolding* (MVU) method based on the notion of local isometry, the MVU method can preserve the original neighborhood relationship of A and B, including the lengths of edges and the angles between edges at the same node (Weinberger and Saul 2004).

DOI: 10.1201/9781003139416-4

After unfolding the input data, the real distance between instances can be truly represented by the unfolded distance (see Figure 4.1b). For this reason, this chapter presents the MVU-based CL-DT method, which can better take advantage of co-location relationship.

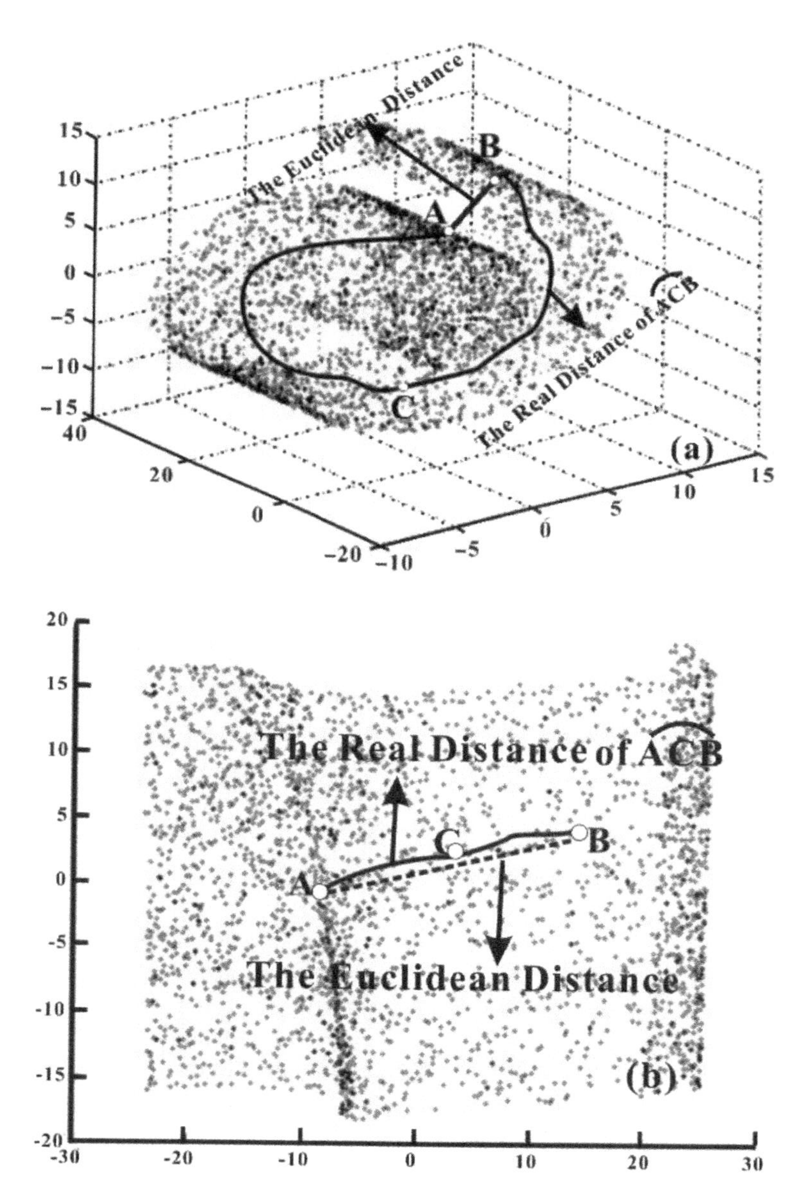

FIGURE 4.1 (a) Data of Swiss roll; (b) the result of mapping the Swiss roll data by MVU.

Note: Figure 4.1 is based on the original figure of Weinberger and Saul 2006.

4.2 MVU-BASED CO-LOCATION PATTERN MINING

4.2.1 Brief Review of MVU

The MVU algorithm was first proposed in 2006 by Weinberger and Saul, whose goal was to detect and discover faithful low dimensional structure in high dimensional data (Weinberger and Saul 2006; Jin et al. 2015; Kanazawaa and Kanatani 2004; Quinlan 1987, 1993). It is based on this intuition: imagine the inputs as connected to their k nearest neighbors by rigid rods. MVU attempts to pull the inputs apart, maximizing the sum total of their pairwise distances without breaking (or stretching) the rigid rods that connect nearest neighbors. The outputs are obtained from the final state of this transformation.

The details of the MVU method can be referenced to Weinberger and Saul (2006). Briefly, the MVU algorithm can faithfully preserves distances and angles among with nearby input data instances by computing the low dimensional representation of a high dimensional data set (Weinberger and Saul 2006; Kardoulas et al. 1996). After MVU processing, the neighbor instances of input and output are invariant about the translation and rotation. Consequently, by calculating the unfolded distances between instances, the real distances between instances can be obtained.

Over the past few decades, many efforts have been made on developing the MVU method. Weinberger and Saul (2004, 2006) put forward a maximum variance unfolding (MVU) algorithm that added a kernel matrix into the algorithm and used the positive semidefinite constraint of the kernel matrix to realize the convex optimization of data. Hou et al. (2008) proposed a relaxed MVU (RMVU) method, which solved the problem that MVU could not unfold manifolds when short-circuit edges appear or when the embedded mapping was conformal but not isometric. Shao and Rong (2009) discussed the deficiencies of kernel principal component analysis for monitoring nonlinear processes and proposed a new process monitoring technique based on maximum variance unfolding projections (MVUP). Liu et al. (2014) analyzed the shortcomings of MVU for process monitoring and proposed an extended maximum variance unfolding (EMVU) method for nonlinear process monitoring. Ery and Bruno (2013) analyzed and discussed the convergence of maximum variance unfolding, and they proved that it is consistent when the underlying submanifold is isometric to a convex subset (Sok et al. 2015; Storey and Choate 2004; Zhou et al. 2016).

4.2.2 MVU-Based Co-Location Pattern Mining

The basic idea of the MVU-based CL-DT method is that the MVU algorithm is applied to unfold the input data for obtaining the real distances (i.e., unfolded distances) between instances. And these unfolded distances are then applied to determine the R-relationship (RRS), which describes the neighbor relation between spatial instances (Huang et al. 2004). With the established RRS of instances, co-location rules are generated to induce the decision-tree generation.

The proposed MVU-based CL-DT method consists of a binary tree structure that utilizes a divide-and-conquer strategy to partition the instance space into decision regions (see Figure 4.2). As shown in Figure 4.2, the proposed method contains three parts:

a. *Unfold input data using MVU method.* First of all, in order to obtain the real distance between instances, original input data set X, which is a non-linear distribution in higher dimensional space, is preprocessed using MVU method. After dealing with MVU, a new data set X′, which is regarded as linear distribution, is produced. Then the data set X′ is employed in two ways: on one hand, the data set X′ is used to calculate unfolding distance; on the other hand, the data set X′ is put into the root node of DT as the input data.
b. *Mine co-location rules.* Co-location rules can be mined by co-location algorithm with the unfolded distances that are calculated in the data set X′.
c. *Generate MVU-based CL-DT.* The data set X′ is also put into the root node of DT as input data. At the beginning, all attributes of data set X′ are "accepted" by the root node, and the DT starts. One "best" attribute, BA_1, is selected from the input data set X′, and then a splitting criterion is used to determine whether the root node will be split using a binary decision (*Yes* or *No* in Figure 4.2) with two intermediate nodes, noted by w_i, and w_j ($w = A$, $i = 1$, and $j = 2$ in Figure 4.2). For each of the intermediate nodes (w_i and w_j), the splitting criterion will be applied to determine whether the node (such as w_i or w_j) should be further split. If *No*, this node is considered a leaf node, and one of the class labels is assigned to this leaf node. If *Yes*, this node will be split by selecting another "best" attribute, BA_2. At the same time, the MVU-based co-location criterion will be used to judge whether the "best" attribute BA_2 is co-located with BA_1 at the same layer. If *Yes*, this node will be "merged" into those nodes with a co-location node. For example, node A1 and A2 are merged into node A1′. After that, one new "best" attribute will be selected again, rejudging whether the selected "best" attribute co-occurs with the last "best" attribute. If *No*, the node will further be split into a subset by repeating this work. This selection process is repeated until a non-co-occurrence attribute is found (Vincent et al. 2010; Wu et al. 1975; Yang 1990).

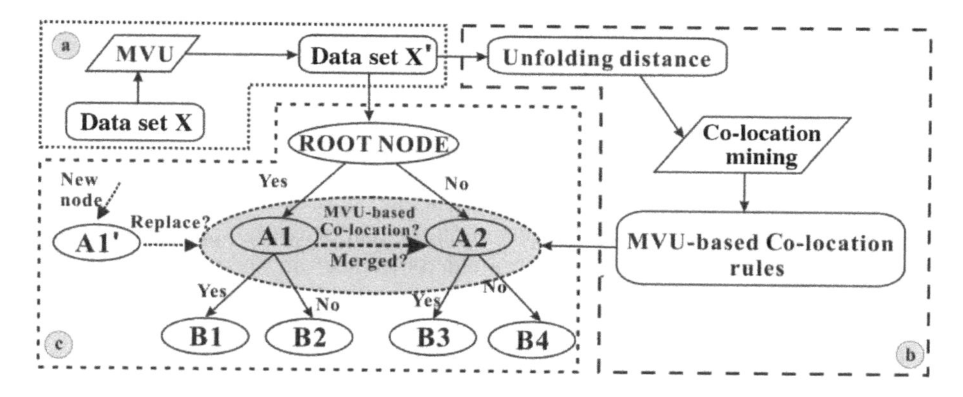

FIGURE 4.2 MVU-based CL-DT induction method.

4.2.3 MVU-Based Co-Location Mining Rules

4.2.3.1 MVU Unfolded Distance Algorithm

The basic idea of MVU unfolded distance algorithm is that every sample instance constitutes a neighborhood relationship matrix with its *k*-nearest instances. If one type of mapping method is found to maximize the distance between two instances that are not nearest neighbors, the embedding of sample instances in the low dimensional space can be realized by using this type of mapping relationship. As a result, the unfolded distance between instances can be calculated and preserved. Thus, the MVU unfolded distance algorithm consists of the three steps: {a) neighbor relation matrix reservation; {b) MVU function establishment and solution; and {c) the calculation of the unfolded distance between instances. They are detailed as follows.

4.2.3.1.1 Neighbor Relation Matrix Reservation

The mapping function for the MVU method is requested to preserve the neighborhood relationship of instances, including the angles and distances of instances. For this reason, it is necessary to set up a matrix to preserve the neighborhood relation. Let $X = \left\{\vec{X}_i\right\}_{i=1}^{N}$ and $Y = \left\{\vec{Y}_i\right\}_{i=1}^{N}$ denote the input data set and output data set, respectively, which have an exactly one-to-one correspondence. We define a matrix δ to preserve the neighborhood relation among instances for X (and similarly, for Y) if input instances X_i and X_kXk (i≠k) are *k*-nearest neighbors, $\delta_{ik} = 1$, or, if X_i and X_kXk (i≠k) are not *k*-nearest neighbors, but another input instance X_j is embedded between X_i and X_k, which makes X_i and X_kXk *k*-nearest neighbors, $\delta_{ik} = 1$ as well; otherwise, $\delta_{ik} = 0$ (and similarly for Y_i and Y_k).

4.2.3.1.2 MVU Function Establishment and Solution

The MVU algorithm "unfolds" the input data by maximizing the sum of pairwise squared distances among output data. Therefore, the objective function can be expressed by (Weinberger and Saul 2006; Breiman et al. 1984; Qin and Karniell 2001):

$$Max\Psi = Max\left\{\sum\nolimits_{i,k}\left\|\vec{Y}_i - \vec{Y}_k\right\|^2\right\} \tag{4.2}$$

where $\left\|\vec{Y}_i - \vec{Y}_k\right\|^2$ is the squared distance of output instances $\vec{Y}_i$ and $\vec{Y}_k$. Because the MVU is a constrained optimization algorithm, it sets up three constraint conditions as follows.

Constraint 1: Local isometry. Based on the characteristics of local isometry before and after mapping, the constraint condition for reserving the angles and distances among *k*-nearest neighbors can be expressed by

$$\delta_{ik}\left|\vec{Y}_i - \vec{Y}_k\right|^2 = \delta_{ik}\left|\vec{X}_i - \vec{X}_k\right|^2 \tag{4.3}$$

Supposing that a squared distance matrix is defined as $D = (D_{ik})_{n \times n}$, that is,

$$D_{ik} = \left| \vec{X}_i - \vec{X}_k \right|^2 \tag{4.4}$$

equation 4.3 can be further expressed by

$$\delta_{ik} \left| \vec{Y}_i - \vec{Y}_k \right|^2 = \delta_{ik} D_{ik} \tag{4.5}$$

Constraint 2: Output centralization. In addition, to eliminate the output instances' translational degree of freedom, which results in the changes of the instances' locations, it is necessary to set up a constraint to ensure that the barycenter of the output data set is in the center. This constraint is mathematically expressed by

$$\sum_i \vec{Y}_i = 0 \tag{4.6}$$

iYi = 0

With this description, the MVU algorithm is a constrained optimal problem. Under the condition of satisfying the constraints of local isometry and output centralization, it maximizes the distance between instances. It is worth noting that the constrained optimal problem is a maximization problem of quadric form under the equality constraints. Thus, it is not a convex optimization problem. For this reason, an inner product matrix L is defined as

$$L_{ik} = \vec{Y}_i \cdot \vec{Y}_k \tag{4.7}$$

equation 4.5 is rewritten by

$$\delta_{ik} \left(L_{ii} - 2L_{ik} + L_{kk} \right) = \delta_{ik} D_{ik} \tag{4.8}$$

and the output centralization can be deduced as

$$0 = \left\| \sum_i \vec{Y}_i \right\|^2 = \sum_i \vec{Y}_i \bullet \vec{Y}_k = \sum_{ik} L_{ik} \tag{4.9}$$

Constraint 3: Symmetric positive semidefinite. Because equation 4.8 and equation 4.9 are linear constraints about L_{ik}, they can be regarded as functions of the inner product matrix L. But not all matrices can be converted into inner product matrices; only symmetric positive semidefinite matrices can undergo such a transformation. A constraint thus must be added to ensure that a matrix is symmetric, positive, and semidefinite. The additional constraint condition can be expressed by

$$L \geq 0 \tag{4.10}$$

Following this new constraint, a new objective function is rewritten as the function about the inner product matrix L, that is,

$$\begin{aligned} Max\Psi &= Max\left\{\sum\nolimits_{i,k}\left\|\vec{Y}_i-\vec{Y}_k\right\|^2\right\} \\ &= Max\left\{\sum\nolimits_{i,k}\left\|\vec{Y}_i\right\|^2\right\} \\ &= Max\left\{\sum\nolimits_{i,k}L_{ii}\right\} \\ &= Max\left\{tr(L)\right\} \end{aligned} \tag{4.11}$$

In combination with equation 4.5, equation 4.6, and equation 4.10, the MVU method can be mathematically expressed by

$$\begin{cases} Max\{tr(L)\} \\ s.t.: \quad (1)\,\delta_{ik}\left(L_{ii}-2L_{ik}+L_{kk}\right)=\delta_{ik}D_{ik} \\ \quad\quad\quad (2)\sum\nolimits_{ik}L_{ik}=0 \\ \quad\quad\quad (3)\,L\geq 0 \end{cases} \tag{4.12}$$

With the MVU model established earlier, positive semidefinite programming (SDP) is employed to solve it. The details can be referenced in Vandenberghe and Boyd (1996). After solution, the inner product matrix L can be obtained, and then the output instance $\vec{Y}_i$ can be obtained by utilizing the method of diagonalization of matrices. This algorithm can be described as follows.

Let $U_{\beta i}$ represents the β-th element of the i-th eigenvector, with eigenvalue $\lambda\beta$; then the inner product matrix L can be rewritten by

$$L_{ik}=\sum\nolimits_{\beta=1}^{n}\lambda_\beta U_{\beta i}U_{\beta k} \tag{4.13}$$

$$Lik=\beta=1n\lambda\beta U\beta iU\beta k$$

Combining equation 4.13 with equation 4.7

$$L_{ik}=Y_i\cdot Y_k,$$

the βth element of output instance $\vec{Y}_i$ is expressed by

$$\vec{Y}_{\beta i}=\sqrt{\lambda_\beta}U_{\beta i} \qquad Y\beta i=\ldots.\lambda\beta U\beta i \tag{4.14}$$

4.2.3.1.3 The Calculation of the Unfolded Distance between Instances

Based on the preserved neighbor relation matrix knowledge, we can use the established MUV model to unfold input data, and then the unfolded distances between instances are calculated. The calculation process can be described as follows. If Y_i

and Y_j are k nearest neighbors, then the unfolded distance, noted as $D_U(Y_i,Y_j)$, can be replaced by the Euclidean distance of Y_i and Y_j, represented as $D_E(Y_i,Y_j)$. Otherwise, if they are not k nearest neighbors, $D_U(Y_i,Y_j)$ is represented by the shortest cumulative Euclidean distance of a number of neighbors. This criterion can mathematically be described by

$$D_U\left(Y_i,Y_j\right)=\begin{cases} D_E\left(Y_i,Y_j\right) \\ min\left\{D_U\left(Y_i,Y_j\right),D_U\left(Y_i,Y_k\right)+D_U\left(Y_k,Y_h\right)+\cdots+D_U\left(Y_g,Y_j\right)\right\} \end{cases} \quad (4.15)$$

$$\textit{if } Y_i \textit{ and } Y_j \textit{ are neighbor}$$

$$\textit{others}$$

To further explain the unfolded distance between instances, an example is given as follows. As shown in Figure 4.3, k nearest neighbors are connected by lines, such as Y_1, Y_2, and Y_3. In Figure 4.3, two cases are considered: (1) the calculation of unfolded distance between k nearest neighbors; (2) and the calculation of unfolded distance between instances that are not k nearest neighbors. For k nearest neighbors, Y_1 and Y_3, according to equation 4.15, the unfolded distance is the Euclidean distance between them. However, for Y_{10} and Y_1, Y_{10} is not one of the k nearest neighbor of Y_1. Hence, on the basic of equation 4.15, the $D_U(Y_1,Y_{10})$ is the shortest cumulative Euclidean distance of a number of neighbors (such as Y_1, Y_3, Y_5, Y_6, Y_9, and Y_{10}). Because output data of MVU can be seen as a connected graph, such a shortest path always can be found.

With this analysis, the MVU unfolded distance (MUD) algorithm is described as follows (Algorithm 4.1).

4.2.3.2 Determination of MVU-Based Co-Locations Patterns

When these steps are completed successfully, only the unfolded distance between instances can be obtained. Based on the unfolded distances, an *RRS* between instances, which describes the neighborhood relation between spatial instances (Chen et al. 2014; Huang et al. 2004; Zhou and Wang 2012), can be created as follows.

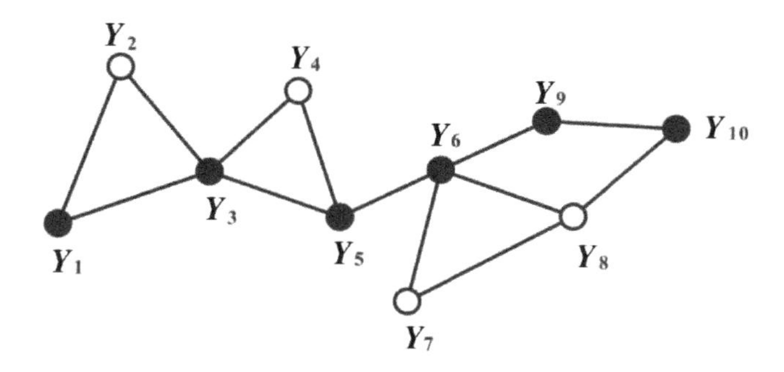

FIGURE 4.3 Determination of the unfolded distance between instance and instance.

Algorithm 4.1 / The Algorithm of Mud /

Input:

1. The number of nearest neighbors k
2. Original data set X
3. An $N \times N$ zero matrix δ

Output:

4. Low dimensional representation Y
5. An $N \times N$ binary matrix δ'
6. An unfolding distance matrix U

Process:

Step 1: Construct neighbor graph.

7. If X_i and X_k satisfy $K - NN$
8. Then $\delta'_{ik} = 1$ else if $\delta'_{ik} = 0$;

Step 2: Semidefinite programming.

9. Compute the maximum variance unfolding Gram matrix L of sample pair-wise points, which is centered on the origin.

$$\text{Max}\{\text{tr}(L)\} \text{ s.t. } L \geq 0, \sum_{ik} L_{ik} = 0$$

$$\text{And } \delta_{ik}\,(L_{ii} - 2L_{ik} + L_{kk}) = \delta_{ik} D_{ik}.$$

Step 3: Compute low dimensional embedding.

10. Perform generalized eigenvalue decomposition for Gram matrix L obtained by Step 2. The eigenvectors, corresponding to the front d greater eigenvalues, is the result of embedding.
11. $\vec{Y}_{\beta i} = \sqrt{\lambda_\beta} U_{\beta i}$

Step 4: Calculate the unfolding distance of instances Y_i and Y_j.

12. $$D_U\left(Y_i, Y_j\right) = \begin{cases} D_E\left(Y_i, Y_j\right) \\ min\left\{D_U\left(Y_i, Y_j\right), D_U\left(Y_i, Y_k\right) + D_U\left(Y_k, Y_h\right) + \cdots + D_U\left(Y_g, Y_j\right)\right\} \end{cases}$$

if Y_i and Y_j are neighbor

others

If and only if the unfolded distance of Y_i and Y_j is less than or equal to the distance threshold, there exists *RRS* between them. The *RRS* is mathematically expressed by

$$R\left(Y_i, Y_j\right) = \begin{cases} 1 & \text{if } D_U\left(Y_i, Y_j\right) \leq D_\theta \\ \text{NAN} & \text{if } D_U\left(Y_i, Y_j\right) > D_\theta \end{cases} \quad (4.16)$$

where D_θ is the threshold of unfolded distance, $R(Y_i, Y_j)$ represents the *RRS* of Y_i and Y_j, and $D_U(Y_i, Y_j)$ is the unfolded distance of Y_i and Y_j. After the determination of the *RRS*, those which satisfy the *RRS* condition are the candidates of MVU-based co-location.

After this processing, only the 2nd-order candidates of MVU-based co-location are obtained. To determine MVU-based co-locations, different attributes (such as surface soil moisture (SSM), land surface temperature (LST), and vegetation coverage (VC)) in the image classification of every instance are utilized. For this reason, a density ratio (*DR*) is defined to determine whether instances are MVU-based co-location patterns. The mathematical model of *DR* is expressed by

$$DR = \frac{similar\_AttrN\left(Y_i, Y_j\right)}{Total\_AttrN} \quad (4.17)$$

where Y_i is the *i*-th instance, $similar\_AttrN\ (Y_i, Y_j)$ is the number of attributes in which the values of Y_i and Y_j satisfy the same threshold, and $Total\_AttrN$ is the total number of attributes used in this chapter. After calculation, if *DR* is greater than or equal to the threshold DR_θ, (Y_i, Y_j) is an MVU-based co-location pattern. For example, in Figure 4.4, there are *RRS*s between arbitrary two instances of A_1, A_6, and A_{11}. (A_1, A_6) is a 2nd-order candidate of a MVU-based co-location pattern, and the total number of utilized attributes, including the components of PCA, VC, LST, SSM, and texture, is five. However, only four attribute (PCA components, LST, SSM, and texture) values of instances A_1 and A_6 satisfy the corresponding attribute threshold, so the *DR* of (A_1, A_6) is 4/5, which is greater than the threshold DR_θ set as 3/5. Thus, (A_1, A_6) is a 2nd-order MVU-based co-location pattern.

When the 2nd-order MVU-based co-location patterns are determined, the 3rd-order and higher order MVU-based co-location patterns should be constructed. The method for constructing the 3rd-order MVU-based co-location patterns is described as follows.

Let the $(k-1)$th order MVU-based co-location pattern C_{k-1} connect the $(k-2)$ th order MVU-based co-location pattern C_{k-2} to generate the *k*-th order candidate of MVU-based co-location pattern C'_k, and if C'_k satisfies the threshold of *DR*, it is regarded as *k*-order MVU-based co-location pattern C_k. Taking the constructed process of the 3rd-order MVU-based co-location (A_1, A_6, A_{11}) in Figure 4.4 as an example. First, (A_1, A_6) connects (A_6, A_{11}) according to the previous $k-2$ order. Then, the candidate of the MVU-based co-location pattern (A_1, A_6, A_{11}) is obtained. Finally, if the *DR* of (A_1, A_6, A_{11}) satisfies the threshold constraint condition, (A_1, A_6, A_{11}) is a MVU-based co-location pattern. By parity of reasoning, all MVU-based co-location patterns will be obtained.

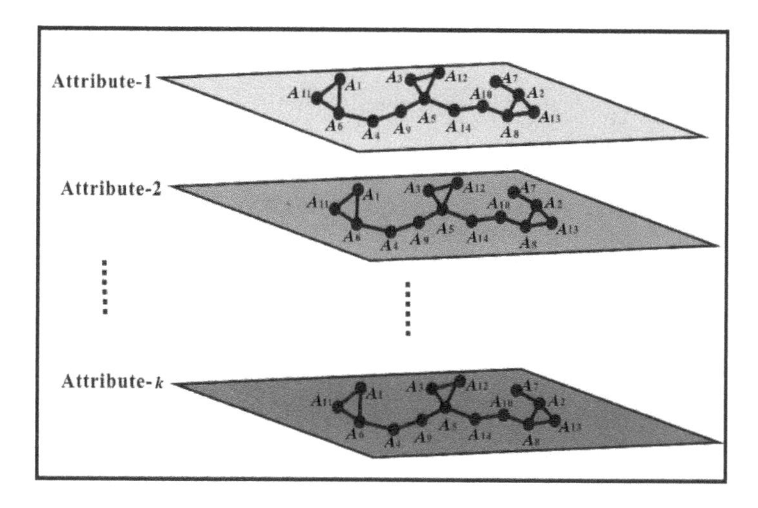

FIGURE 4.4 Determination of MVU-based co-locations (Note: Attribute-*k* represents the *k*-th attribute, and A_i represents the *i*-th instance of object *A*. In each of the attributes, the location of instance A_i is the same).

Because those instances that are MVU-based co-location patterns have similar characteristics, the phenomenon that different objects have the same spectrum in remote sensing images will be eliminated. Thus, the classification accuracy will be enhanced.

4.2.3.3 Determination of Distinct Event-Types

These processes are used only to mine MVU-based co-location instances but cannot ensure that these instances are distinct event types. Therefore, a constraint condition has to be set.

Let $M = \{m_1, m_2, \ldots, m_c\}$ be a set of corresponding cluster centers of attributes $a_1, a_2, \ldots, a_c$, so the distinct event-type can be expressed by

$$\Psi_i = \sum_{i=1}^{S} \sum_{k=1}^{C} \left(\| f_i - m_k \| \right)^2 \tag{4.18}$$

where $\|f_i - m_k\|$ denotes the Euclidean distance between f_i and m_k; f_i is the value of the *i*-th instance in attribute a_k; Ψ_i is a squared error clustering criterion; *C* is the number of attributes; and *S* is the number of instances. Therefore, if Ψ_i is greater than the given threshold Ψ_θ, then the *i*-th instance is regarded as a distinct event.

4.2.4 Generation of MVU-Based Co-Location Rules

Accompanying by the generation of MVU-based co-locations, the MVU-based co-location rules can be generated from the candidates of MVU-based co-location. On the basic of the previous analysis, the algorithm of determination of MVU-based co-locations (MCL) developed in this paper is summarized, and the pseudocodes are shown in Algorithm 4.2.

Algorithm 4.2 / The Algorithm of Determination of Mcl /

Input:

1. Unfolded distance matrix D_U;
2. Unfolded data set Y;
3. Thresholds of distance D_θ, and thresholds of density radio $\Re_\theta$;

Valuable:

4. ζ: the order of co-location pattern;
5. R: the set of R-relationship between instances;
6. C_ζ: the set of candidates of co-location pattern whose size is ζ;
7. Tab_ζ: the set of co-location patterns;
8. Rul_ζ: the ζ order co-location rule.

Output:

9. ζ order co-location pattern; Rul_ζ.

Process:

Step 1:

10. If $(D_U(Y_i, Y_j) \leq D_\theta)$ Then $(R(Y_i, Y_j) = 1)$
11. else $(R(Y_i, Y_j) = NAN)$;

Step 2:

12. MVU_co-location size $\zeta = 1$; $C\_1 = Y_i$; $P\_1 = Y_i$;

Step 3:

13. $Tab\_1 = generate\_table\_instance\ (C\_1, P\_1)$;

Step 4:

14. If (fmul = TRUE) Then
15. $Tab\_C\_1 = generate\_table\_instance\ (C\_1, multi\_event)$;

Step 5:

16. $R = determine\_R\text{-}relationship\ (D_U(Y_i, Y_j))$;
17. While (P_ζ is not empty and $\zeta < k$) do
18. $\{C\_\zeta + 1 = generate\_candidate\_MVU\text{-}co\text{-}location\ (C\_\zeta, \zeta)$
19. $Tab\_C\_\zeta + 1 = generate\_MVU\text{-}co\text{-}location\ (\Re_\theta, C\_\zeta, \Re)$
20. $Rul\_\zeta + 1 = generate\_MVU\text{-}co\text{-}location\_rule\ (Tab\_C\_\zeta + 1, \beta)$
21. $\zeta{+}{+}\}$
22. Return ζ order co-location pattern; Rul_ζ.

4.3 PRUNING

In a DT induction algorithm, different pruning algorithms are usually employed to prune the nodes. In this chapter, a pruning algorithm proposed by Zhou and Wang (2012) is employed. The pruning algorithm is briefly described as follows.

For a spatial data set *SP*, let $FT = (ft_1, ft_2, ft_3, \cdots, ft_h)$ be a set of spatial attributes. Let $\Phi = (\varphi_1, \varphi_2, \varphi_3, \cdots, \varphi_n)$ be a set of n instances in *SP*, where each instance is a vector instance-ID, spatial attributes, and location. The spatial attribute of instance i is denoted by ft_i. It is assumed that the spatial attributes of an instance are from spatial attribute set *FT*, that the location is within the spatial framework of a spatial database, and that there is an RRS in *SP*. Additionally, let $O = (o_1, o_2, o_3, \ldots, o_k)$ be a set of corresponding clusters centered in the data set *SP*, where k is the number of clusters of spatial attributes. To capture the concept of "nearby", the criterion of co-occurrence is defined as

$$\Xi_m = \sum_{i=1}^{h} \sum_{q=1}^{n} v_{iq} \left(\left\| \varphi_q - o_i \right\| \right)^2 \tag{4.19}$$

where $\|\varphi_q - o_i\|$ is the Euclidean distance between x_q and o_i; Ξ_m is the squared error clustering criterion; and $V = \{v_{iq}\}$, $i = 1, 2, \ldots, h$, $q = 1, 2, \ldots, N$ is a matrix that satisfies the following constraint conditions:

$$v_{iq} \in [0,1], \quad \forall_i = 1,2,\cdots,h, \quad \forall q = 1,2,\cdots,N \tag{4.20}$$

$$\sum_{i}^{h} v_{iq} = 1, \quad \forall_i = 1,2,\cdots,h, \quad \forall q = 1,2,\cdots,N \tag{4.21}$$

So, if Ξ_m is less than or equal to a given threshold, the two nodes are considered as co-occurrent and thus should be merged.

4.4 INDUCTING DECISION RULES

After the MVU-based CL-DT is generated, decision rules will be created by translating the decision tree into semantic expressions. Because the MVU-based CL-DT algorithm partitions a data space into several distinct disjoint regions via axis parallel surfaces, the top-down search method will be employed to translate individual node into rules in this chapter.

In a nutshell, the whole flowchart of the proposed MVU-based CL-DT algorithm is shown in the following chart (see Algorithm 4.3).

Algorithm 4.3 / The Algorithm of Mvu-Based Cl-Dt/

Input:

1. Training data set *TD*; the threshold of unfolded distance D_θ;
2. The thresholds of density radio $\Re_\theta$; splitting criterion;
3. The threshold of terminal node;

Output:

4. A MVU-based CL-DT with multiple condition attributes.

Process:

Step 1: The algorithm of MUD: function MUD;
Step 2: The algorithm of determination of MCL: function MCL;
Step 3: Judge whether co-locations are distinct event type;
Step 4: Build an initial tree;
Step 5: Starting with a single node, root, which includes all rules

5. and attributes;

Step 6: Judge whether each non-leaf node will be further split,

6. e.g., w_i;
7. • Perform label assignment test to determine if there are any
8. labels that can be assigned;
9. • An attribute according to splitting criterion is selected to split
10. w_i further and judge whether stop criterion is meet;
11. O If the selected attribute meet the splitting criterion, the node
12. will be parted into a subset;
13. O If the stop criterion is satisfied, stop splitting and assign w_i
14. as a leaf node;

Step 7: Apply MVU-based co-location algorithm for each of two

15. non-leaf nodes in the same layer, e.g., w_i and w_j, to test
16. whether the two nodes satisfy the co-location criterions. If
17. yes, merging the two neighbor nodes; if no, go back to Step 6.

Step 8: Apply the algorithm recursively to each of the not-yet-

18. stopped nodes;

Step 9: Generate decision rules by collecting decisions driven in

19. individual nodes;

Step 10: The decision rules generated in Step 7 are used as the

20. initialization of co-location mining rule and apply the
21. algorithm of the co-location mining rule to generate new
22. associate rules.

Step 11: Reorganize the input data set, and repeat Step 2 through

23. Step 8 until the classified results from the colocation mining
24. rule and decision tree (rules) are consistent.

REFERENCES

Appel, R., et al., Quickly boosting decision trees: Pruning underachieving features early. *Proceedings of the 30th International Conference on Machine Learning*, vol. 28, Atlanta, GA, USA, June 16–21, 2013, pp. 594–602.

Breiman, L., Friedman, J.H., Olshen, R.A., and Stone, C.J., *Classification and regression trees*. Belmont, CA, USA: Wadsworth, Chapman and Hall/CRC, 1984. ISBN-13: 978-0412048418.

Chasmer, L., et al., A decision-tree classification for low-lying complex land cover types within the zone of discontinuous permafrost. *Remote Sensing of Environment*, vol. 143, 2014, pp. 73–84.

Chen, Y., et al., Deep learning-based classification of hyperspectral data. *IEEE Journal of Selected Topics in Applied Earth Observations and Remote Sensing*, vol. 7, no. 6, 2014.

Ery, A.C., and Bruno, P., On the convergence of maximum variance unfolding. *Journal of Machine Learning Research*, vol. 14, 2013, pp. 1747–1770.

Farid, D., et al., Hybrid decision tree and naïve Bayes classifiers for multi-class classification tasks. *Expert Systems with Applications*, vol. 41, 2014, pp. 1937–1946.

Franco-Arcega, A., and Carrasco-Ochoa, J.A., Sanchez-Diaz, G., and Martinez-Trinidad, J.F., Decision tree induction using a fast splitting attribute selection for large datasets. *Expert Systems with Applications*, vol. 38, 2011, pp. 14290–14300.

Franklin, S.E., et al., An Integrated Decision Tree Approach (IDTA) to mapping land cover using satellite remote sensing in support of grizzly bear habitat analysis in the Alberta yellow head ecosystem. *Canadian Journal of Remote Sensing*, vol. 27, 2001, pp. 579–592.

Greiner, G., and Hormann, K., Efficient clipping of arbitrary polygons. *ACM Transactions on Graphics*, vol. 17, no. 2, April 1998, pp. 71–83.

Hou, C.P., et al., Relaxed maximum-variance unfolding. *Optical Engineering*, vol. 47, no. 7, July 17, 2008. pp. 077202–1–12.

Huang, X., Zhang, L., and Wang, L., Evaluation of morphological texture features for mangrove forest mapping and species discrimination using multispectral IKONOS imagery. *IEEE Geoscience and Remote Sensing Letters*, vol. 6, no. 3, June 2009, pp. 393–397.

Huang, Y., Shekhar, S., and Xiong, H., Discovering co-location patterns from spatial data sets: A general approach. *IEEE Transactions on Knowledge and Data Engineering*, vol. 16, no. 12, 2004, pp. 1472–1485.

Jin, J., Zheng, Y., and Xin, D., Selection of the optimal decision tree model using grid search method: Focusing on the analysis of the factors affecting job satisfaction of workplace reserve force commanders. *Journal of the Korean Operations Research and Management Science Society*, vol. 40, no. 2, 2015, pp. 19–29.

Kanazawaa, Y., and Kanatani, K., Image mosaicing by stratified matching. *Image and Vision Computing*, vol. 22, 2004, pp. 93–103.

Kardoulas, N.G., Bird, A.C., and Lawan, A.I., Geometric correction of spot and Landsat imagery: A comparison of map-and GPS-derived control points. *American Society for Photogrammetry and Remote Sensing*, vol. 62, no. 10, 1996, pp. 1171–1177.

Lacar, F.M., Lewis, M.M., and Grierson, I.T., Use of hyperspectral imagery for mapping grape varieties in the Barossa Valley, South Australia. *Proceedings of IEEE Int. Geoscience and Remote Sensing Symposium (IGARSS)*, vol. 6, Sydney, Australia, 2001, pp. 2875–2877.

Liu, Y.J., Chen, T., and Yao, Y., Nonlinear process monitoring and fault isolation using extended maximum variance unfolding. *Journal of Process Control*, vol. 24, 2014, pp. 880–891.

Mansour, Y., Pessimistic decision tree pruning based on tree size. *Proceedings of 14th International Conference on Machine Learning*, San Francisco, CA, USA, 1997, pp. 195–201.

Mohammad, A., Shi, Z., and Ahmad, Y., Application of GIS and remote sensing in soil degradation assessments in the Syrian coast. *Journal of Zhejiang University (Agric. & Life Sci.)*, vol. 26, no. 2, 2002, pp. 191–196.

Moustakidis, S., et al. SVM-based fuzzy decision trees for classification of high spatial resolution remote sensing images. *IEEE Transactions on Geoscience and Remote Sensing*, vol. 50, no. 1, 2012, pp. 149–169.

Osei-Bryson, K., Post-pruning in decision tree induction using multiple performance measures. *Computers & Operations Research*, vol. 34, no. 11, November 2007, pp. 3331–3345.

Osei-Bryson, K., Post-pruning in regression tree induction: An integrated approach. *Expert Systems with Applications*, vol. 34, 2008, pp. 1481–1490.

Polat, K., and Gunes, S., A novel hybrid intelligent method based on C4.5 decision tree classifier and one-against-all approach for multi-class classification problems. *Expert Systems with Applications*, vol. 36, 2009, pp. 1587–1592.

Qin, Z., and Karniell, A., A mono-window algorithm for retrieving land surface temperature from Landsat TM data and its application to the Israel-Egypt border region. *International Journal of Remote Sensing*, vol. 22, no. 18, 2001, pp. 3719–3746.

Quinlan, J.R., *C4.5: Programs for machine learning*. San Mateo, CA: Morgan Kaufmann, 1993. ISBN: 978-1-55860-238-0.

Quinlan, J.R., Induction of decision trees. *Machine Learning*, vol. 1, 1987, pp. 81–106.

Shao, J.D., and Rong, G., Nonlinear process monitoring based on maximum variance unfolding projections. *Expert Systems with Applications*, vol. 36, 2009, pp. 11332–11340.

Simard, M., Saatehi, S.S., and Grandi, G.D., Use of decision tree and multi-scale texture for classification of JERS-1 SAR data over tropical forest. *IEEE Transactions on Geoscience and Remote Sensing*, vol. 38, 2000, pp. 2310–2321.

Sok, H.K., Ooi, M.P., and Kuang, Y.C., Sparse alternating decision tree. *Pattern Recognition Letters*, vol. 60–61, 2015, pp. 57–64.

Storey, J.C., and Choate, M.J., Landsat-5 bumper-mode geometric correction. *IEEE Transactions on Geoscience and Remote Sensing*, vol. 42, no. 12, 2004, pp. 2695–2703.

Vandenberghe, L., and Boyd, S.P., Semidefinite programming. *Society for Industrial and Applied Mathematical Review*, vol. 38, no. 1, 1996, pp. 49–95.

Vincent, P., et al., Stacked denoising autoencoders: Learning useful representations in a deep network with a local Denoising criterion. *The Journal of Machine Learning Research*, vol. 11, 2010, pp. 3371–3408.

Weinberger, K., and Saul, L., Unsupervised learning of image manifolds by semidefinite programming. *Proceedings of CVPR 2004*, IEEE Computer Society, Los Alamitos, 2004, pp. 988–995.

Weinberger, K., and Saul, L., Unsupervised learning of image manifolds by semidefinite programming. *International Journal of Computer Vision*, vol. 70, no. 1, 2006, pp. 11–90.

Witten, I.H., and Frank, E., *Data mining-practical machine learning tools and techniques with Java implementation*. San Mateo, CA: Morgan Kaufmann, 2000. ISBN号: 978-1-55860-238-0.

Wu, C., Landgrebe, D., and Swain, P., The decision tree approach to classification. *School Elec. Eng., Purdue University, West. Lafayette, Indiana*, Report, RE-EE 75–17, 1975.

Xu, C.G., and Anwar, A., Based on the decision tree classification of remote sensing image classification method application. *Applied Mechanics and Materials*, vol. 316–317, 2013, pp. 193–196.

Yang, W.J., The registration and mosaic of digital image remotely sensed. *Proceedings of the 11th Asian Conference on Remote Sensing*, vol. 2, ACRSQ-12–1–6, Guangzhou China, November 15–21, 1990, pp. 1–6.

Zhan, D., and Hua, Z., Ensemble-based manifold learning for visualization. *Journal of Computer Research and Development*, vol. 42, no. 9, 2005, pp. 1533–1537.

Zhang, N., Wu, Y., and Zhang, Q., Detection of sea ice in sediment laden water using MODIS in the Bohai Sea: A CART decision tree method. *International Journal of Remote Sensing*, vol. 36, no. 6, 2015, pp. 1661–1674.

Zhang, R., The research on algorithm of MVU based co-location decision tree. *MS. Thesis*, Guilin University of Technology, July 2015.

Zhou, G., Co-location decision tree for enhancing decision-making of pavement maintenance and rehabilitation. *Ph.D. Dissertation*, Virginia Tech, Blacksburg, Virginia, USA, 2011.

Zhou, G., and Wang, L., Co-location decision tree for enhancing decision-making of pavement maintenance and rehabilitation. *Transportation Research Part C: Emerging Technologies*, vol. 21, no. 1, 2012, pp. 287–305.

Zhou, G., Zhang, R., and Zhang, D. Manifold learning co-location decision tree for remotely sensed imagery classification. *Remote Sensing*, vol. 8, no. 10, 2016, p. 855.

5 Maximal Instance Co-Location Pattern Mining Algorithms

5.1 INTRODUCTION

Spatial co-location mining is used to mine the "positive" relationship of different spatial features in spatial data. It is somewhat difficult to find co-location instances, since the instances of spatial features are embedded in a continuous space and share neighbor relationships. Association rule mining algorithms cannot be directly applied to co-location pattern mining since there are no predefined transactions in spatial data sets (Huang et al. 2003, 2004). Thus, the following three basic approaches are proposed for mining co-locations.

Join-based approach: In order to solve the problem of mining co-locations, Huang et al. first proposed a transaction-free (join-based) approach (Huang et al. 2004). This approach mines co-locations by using the concept of proximity neighborhood and identifies co-location instances by joining table instances, which ensure the correctness and completeness of this approach. However, it is time-consuming due to a large number of join operations required as the number of features and their instances increases. Figure 5.1b shows the detailed process of instance join. Huang et al. addressed the problem of mining co-location patterns with rare spatial events (Huang et al. 2006). In their paper, a new measure called the maximal participation ratio (maxPR) was introduced, and a weak monotonicity property of the maxPR measure was identified. Verhein introduced the concept of maximal clique and applied the GLIMIT (Geometrically Inspired Linear Itemset Mining in the Transpose) itemset mining algorithm to mine complex spatial co-location patterns, which leads to far more superior performance than using an Apriori style approach (Verhein and Al-Naymat 2007). Alnaymat used maximal clique for mining co-locations from a Sloan Digital Sky Survey (SDSS) data (Alnaymat et al. 2008); Kim proposed a polynomial algorithm called *AGSMC (Algorithm Generating Spatial Maximal Cliques)* to generate all maximal cliques from general spatial data sets; AGSMC constructs the tree-type data structures using the materializing method and generates maximal cliques by scanning the constructed trees (Kim et al. 2011). Yao et al. propose an adaptive maximal co-location (AMCM) algorithm to address two limitations in the co-location mining method: (1) it is difficult to set an appropriate proximity threshold to identify close instances in an unknown region and (2) those methods neglect the effects of distance values between instances and far-instance effects on the pattern's significance (Yao et al. 2017). Deng et al. developed a multi-level method to identify regional co-location patterns in two steps. First, global co-location patterns

DOI: 10.1201/9781003139416-5

were detected and other non-prevalent co-location patterns were identified as candidates for regional co-location patterns (Deng et al. 2017). Second, an adaptive spatial clustering method was applied to detect the subregions where regional co-location patterns are prevalent.

Partial-join approach: Yoo and Shekhar first proposed a partial-join approach, since a large fraction of the join-based co-location miner algorithm is devoted to computing joins to identify instances of candidate co-location patterns (Yoo et al. 2004, 2005, 2014). The partial-join approach first transactionizes continuous spatial data (builds cliques) to identify the intraX instances of co-location (belonging to a clique) and interX instances of co-location (belonging between two cliques) then joins the intraX instances and interX instances. respectively. to calculate the value of the participation index. This approach mines co-location patterns more efficiently than the join-based approach since it reduces a large number of join operations. However, building cliques is time consuming. Huang et al. proposed that spatial clustering groups similar spatial objects together and proposed a new approach to the problem of mining co-location patterns using clustering techniques (Huang et al. 2008). Yu proposed a new co-location analysis approach to find the prevalent regions of a pattern; the approach combines kernel density estimation and polygon clustering techniques to specifically consider the correlation, heterogeneity, and contextual information existing within complex spatial interactions (Yu 2017). Ouyang et al. studied the co-location mining problem for fuzzy objects and proposed two new kinds of co-location pattern mining for fuzzy objects, single co-location pattern mining (SCP) and range co-location pattern mining (RCP), to mining co-location patterns at a membership threshold or within a membership range (Ouyang et al. 2017). Celik et al. proposed an indexing structure for co-location patterns and proposed algorithms (Zoloc-Miner) to discover zonal co-location patterns efficiently for dynamic parameters; extensive experimental evaluation shows their approaches are scalable, efficient, and outperform naive alternatives (Celik et al. 2007).

Join-less approach: In order to solve the problem of excessive time consumption caused by join operations thoroughly, Yoo and Shekhar first proposed a join-less approach (Yoo et al. 2005). The join-less approach uses the star neighborhood to materialize spatial relationships. It is more efficient compared with the join-based approach and partial-join approach since it generates co-location patterns without join operations. With the increase of co-location size, the processes of generating table instances of candidate co-locations and prevalent co-locations are repeated, and the computation time of the join-less approach increases. So the efficiency of the join-less approach is affected by the length of co-location patterns. The concept of the negative co-location patterns was defined by Jiang et al. (2010). Based on the analysis of the relationship between the negative and positive participation index, they proposed methods for negative participation index calculation and negative patterns pruning strategies. Zhou and Wang applied co-location patterns to the decision tree; they developed a co-location decision tree (CL-DT) method (Zhou and Wang 2010). Yu investigated a projection-based co-location pattern mining paradigm (Yu 2005). In particular, an FP-tree based colocation mining framework and an algorithm called FP-CM, for FP-tree based colocation miner, are proposed.

Although the three basic approaches have solved the problem of co-location mining, they still have shortcomings: (1) the join-based approach requires a large number of join operations; (2) building cliques is time-consuming in the partial-join approach; (3) the computation time of the join-less approach increases with the increase of the co-location size. For this reason, this chapter intends to overcome these shortcomings through developing a maximal instance algorithm. This algorithm is designed to find maximal instances from a spatial data set. Relying on the definition of the spatial neighbor relationship in the join-based approach, this chapter constructs a RI-tree to generate maximal instances. The construction of an RI-tree is actually a process of judging the spatial neighbor relationship between spatial instances and the series of spatial instances. The advantage of maximal instance algorithm is that it generates co-locations without requirement of join operations, since all row instances are generated by finding subsets of maximal instances and the process of finding subsets does not require joins between row instances (Adilmagambetov et al. 2013; Albanese et al. 2011; Bao and Wang 2017; Celik et al. 2012, Celtic 2011; Ding et al. 2008; Kim et al. 2014; Lei et al. 2012; Li et al. 2004; Li 2020; Mohan et al. 2011; Qian et al. 2009; Shekhar and Huang 2001; Sundaram et al. 2012; Venkatesan and Thangavelu 2013; Wang et al. 2018; Williams et al. 2006; Xiao 2009; Yang et al. 2011; Yao et al. 2016; Zhang 2000; Zhou et al. 2005; Yoo and Shekhar 2006).

5.2 MAXIMAL INSTANCE ALGORITHMS

In Figure 5.1a, each instance is uniquely identified by $T.i$, where T is the spatial feature type and i is the unique ID inside each spatial feature type, lines between instances represent neighbor relationships. Figure 5.1b shows the instances of co-location $\{A, B, C\}$ being generated by instance joins. A maximal instance is a row instance that does not appear as a subset of another row instance. In other words, maximal instances cannot combine with other instances to generate row instances of higher size co-locations. For example, $\{A.1, B.2\}$ in Figure 5.1a is not a maximal instance, since it is a subset of row instance $\{A.1,\ B.2,\ C.1\}$, which in turn is a

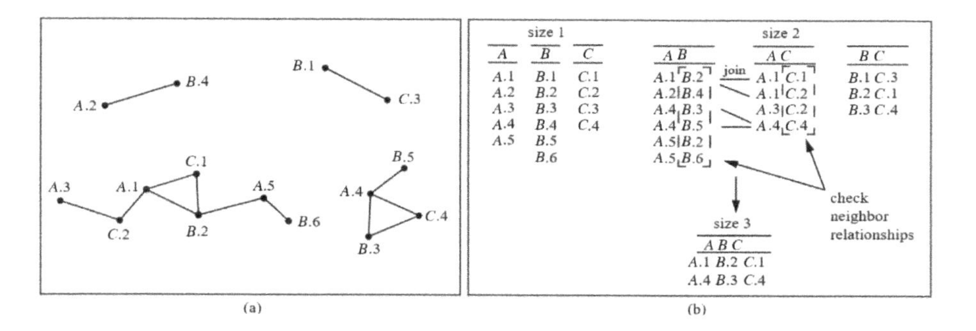

FIGURE 5.1 (a) An example data set. (b) Instance joins.

maximal instance. A maximal instance must be a maximal clique (maximal clique is defined as a clique that does not appear as subset of another clique in the same co-location pattern), but a maximal clique may not be a maximal instance, since maximal cliques may not be row instances. Assuming $A.i$, $B.i$, $B.j$ ($A.i$ is an instance of A, $B.i$ and $B.j$ are instances of B) form a clique, that is, they are neighbors to each other, $\{A.i, B.i, B.j\}$ is a maximal clique rather than a maximal instance as it is not a row instance. However, $\{A.1, B.2, C.1\}$ is a maximal instance and also a maximal clique.

5.2.1 Generation of Row Instances

The generation of row instances is the basic part of co-location mining, because only when all row instances are found can the participation index of co-locations be calculated. How to find row instances of co-locations is discussed here.

In the process of generating row instances of all co-locations, there are some differences between our algorithm and previous algorithms. The generation of row instances in join-based algorithms, based on join operation, can be represented in equation 5.1, and the detailed instance join process for the example data set (in Figure 5.1a) is shown in Figure 5.1b. The k-size row instances are generated from $k-1$-size row instances, that is, high-size row instances are generated by join operations between low-size row instances. For example, 2-row instance $\{A.1, B.2\}$ and $\{A.1, C.1\}$ can be joined to generate 3-row instance $\{A.1, B.2, C.1\}$. This method for mining co-location is time-consuming due to the large number of join operations required as the numbers of features and their instances increases.

$$E(RI_1) \xrightarrow{join} RI_2 \xrightarrow{join} \cdots \xrightarrow{join} RI_k \tag{5.1}$$

Here E is a set of spatial event types (i.e., RI_1) and RI_i is a set of row instances of i-size co-locations. In a join-based algorithm, the set of all row instances RI is a merge of RI_1, RI_2, $\cdots$, and RI_k, that is, $RI = \sum_{i=1}^{k} RI_i$. Every time the size of co-locations is increased, a join operation needs to be performed, resulting in $k-1$ times of join operations needed to generate a k-size row instance. The larger k, the more join operations needed and the more execution time consumed. Therefore, the method (called a join-based method by us) of generating row instances by join operations is inefficient.

By reviewing the generation of row instances in the join-based algorithm, the relationship between a high-size row instance H generated from L with a low-size row instance L is obvious: L is a subset of H since H must contain the instances in L. And if a high-size row instance exists, it must also exist for its corresponding low-size row instances. The corresponding low-size row instances are actually subsets of the high-size row instances. So *Lemma 1* can be obtained. According to *Lemma 1*, instead of using low-size row instances to generate high-size row instances, as in traditional algorithms, this chapter thinks backwards: using high-size row instances to generate low-size row instances.

Lemma 1

An arbitrary non-empty subset of a k*-size row instance must be a row instance of co-locations.*

Proof. This lemma is proved from the definition and generation process of row instances. (1) By definition, a row instance is essentially an **R**-proximity neighborhood. And the definition of R-proximity neighborhoods is given by Huang: a R-proximity neighborhood is a set $\mathrm{I} \subset \mathrm{S}$ of instances that form a clique under the relation R. Accordingly, a set I is a row instance if any two instances in I satisfy the neighborhood relation. (2) Huang, et al. gave a detailed introduction to the generation of row instances (Huang et al. 2004). Similar to the generation of candidate co-locations, row instances are generated by a join strategy. An example in Figure 5.1 is used to illustrate the generation of row instances. Row instances $\{A.1,\ B.2\}$ and $\{A.1,\ C.1\}$ are joined to generate $\{A.1,\ B.2,\ C.1\}$; then check the neighbor relationship between $B.4$ and $C.1$. $\{A.1,\ B.2,\ C.1\}$ is a true row instance if $B.2$ has neighbor relationship with $C.1$. In contrast, $\{A.4,\ B.5,\ C.4\}$, generated by joining $\{A.4,\ B.5\}$ and $\{B.5,\ C.4\}$, fails to be a row instance, since $A.4$ has no neighbor relationship with $C.4$. Thus, any two instances in a row instance satisfy the neighbor relationship. Let I_k be a set of instances in a k-size row instance. Because k-size row instances are generated from k – 1-size row instances, a conclusion is drawn: I_{k-1} is a proper subset of I_k (shown as the formula: $I_{k-1} \subseteq I_k$). And it can be inferred that: $I_1 \subseteq I_2 \subseteq \cdots \subseteq I_k$. So any set $\mathrm{I_n}$ is a proper subset of I_k ($1 \le n < k$), that is, $I_n \subseteq I_k$. If row instances are treated as sets, the following conclusions can be drawn: $RI_n \subseteq RI_k$. Here RI_k is a set of k-size row instances, $1 \le n < k$. In summary, an arbitrary non-empty subset of a k-size row instance must be a row instance of co-locations.

Based on: (1) A maximal instance is the maximal row instance for a co-location; it cannot join with another instance to generate a high size row instance; (2) *Lemma 1*, it can draw a further conclusion that non-empty subsets of maximal instances are all row instances of co-locations. Once all maximal instances are found, all row instances can be generated by finding non-empty subsets of them. This method (shown in equation 5.2; $M = \{m_1,\ m_2,\ \cdots,\ m_k\}$ is a set of maximal instances) for generating row instances does not require join operations, which can reduce a lot of computing time. The set of all row instances RI is a merge of non-empty subsets of maximal instances. Empty sets are not considered here since they are meaningless.

$$E \rightarrow M \xrightarrow{gen\_non-empty\ subset} RI \qquad (5.2)$$

A simple comparison is made to illustrate that this method is more efficient than the join-based method for generating row instance. The join-based method generates row instances by join operations. Two-size row instances are first generated by joining spatial instances that have neighbor relationships. Then high-size row instances are generated size by size until no higher-size row instances are generated. It requires k – 1 times of join operations to generate all row instances, which is time consuming. Our method is first to generate the highest-size row instances (maximal instances) then generate all row instances by finding non-empty subsets of the highest-size row instances, which requires no join operations and only one time of finding subsets. One obvious difference between two methods is that the join-based method generates row instances from low-size to high-size; our method generates row instances from highest-size to low-size. A large number of join operations are not needed in our method, which makes our method less time consuming.

Let k be the number of maximal instances; m_i represents a maximal instance $(1 \leq \text{i} \leq \text{k})$; $NS(m_i)$ represents the set of non-empty subsets of m_i; $\mathfrak{B}(m_i)$ is the power set of m_i; $|m_i|$ is the number of instances in m_i; $|RI|$ is the number of row instances; two lemmas can be derived: $RI = \cup NS(m_i) = (\cup \mathfrak{B}(m_i)) - \varnothing$; $|RI| = \sum_{i=1}^{k} 2^{|m_i|} - k$. The proofs are as follows. It should be noted that the universal set here is the set of spatial instances I.

Lemma 2

$$RI = \cup NS(m_i) = (\cup \mathfrak{B}(m_i)) - \varnothing.$$

Proof. First, according to the conclusion that non-empty subsets of maximal instances are all row instances of co-locations, there is $RI = \cup NS(m_i)$. Second, $NS(m_i) = \mathfrak{B}(m_i) - \varnothing$, since: $\mathfrak{B}(m_i)$, the power set of m_i, is defined as a set of all subsets of m_i; $NS(m_i)$ is the set of non-empty subsets of m_i. Therefore,

$$\begin{aligned} RI &= \cup NS(m_i) \\ &= \cup(\mathfrak{B}(m_i) - \varnothing) \\ &= \cup(\mathfrak{B}(m_i) - \varnothing) \\ &= \cup(\mathfrak{B}(m_i) \cap \sim \varnothing) \\ &= (\cup \mathfrak{B}(m_i)) \cap \sim \\ &= (\cup \mathfrak{B}(m_i)) - \varnothing. \end{aligned} \tag{5.3}$$

Lemma 3

$$|RI| = \sum_{i=1}^{k} 2^{|m_i|} - k.$$

Proof. For an n-set, the total number of subsets is $C_n^0 + C_n^1 + \cdots + C_n^n = 2^n$, so $|\mathfrak{B}(m_i)| = 2^{|m_i|}$. And $|NS(m_i)| = |\mathfrak{B}(m_i)| - 1$, since $NS(m_i) = \mathfrak{B}(m_i) - \varnothing$. Therefore, $|RI| = \sum_{i=1}^{k} |NS(m_i)| = \sum_{i=1}^{k} (|\mathfrak{B}(m_i)| - 1) = \sum_{i=1}^{k} (2^{|m_i|} - 1) = \sum_{i=1}^{k} 2^{|m_i|} - k$.

As shown in Table 5.1, $M = \{\{A.1, B.2, C.1\}, \{B.2, A.5\}, \{A.5, B.6\}, \{A.2, B.4\},$ $A.3, C.2\}, \{C.2, A.1\}, \{A.4, B.3, C.4\}, \{A.4, B.5\}, \{B.1, C.3\}\}$ is a set of maximal instances found from the example data set in Figure 5.1(a). Row instances of all co-locations listed in the second column of Table 5.1 are non-empty subsets of maximal instances. And the number of row instances $|RI|$ can be calculated by *Lemma 3*.

TABLE 5.1
Enumeration of Maximal Instances and Co-Location Instances

M	Row instances of all co-locations
$\{A.1, B.2, C.1\}$	$\{A.1\}$, $\{B.2\}$, $\{C.1\}$, $\{A.1, B.2\}$, $\{A.1, C.1\}$, $\{B.2, C.1\}$, $\{A.1, B.2, C.1\}$
$\{B.2, A.5\}$	$\{B.2\}$, $\{A.5\}$, $\{B.2, A.5\}$
$\{A.5, B.6\}$	$\{A.5\}$, $\{B.6\}$, $\{A.5, B.6\}$
$\{A.2, B.4\}$	$\{A.2\}$, $\{B.4\}$, $\{A.2, B.4\}$
$\{A.3, C.2\}$	$\{A.3\}$, $\{C.2\}$, $\{A.3, C.2\}$
$\{C.2, A.1\}$	$\{C.2\}$, $\{A.1\}$, $\{C.2, A.1\}$
$\{A.4, B.3, C.4\}$	$\{A.4\}$, $\{B.3\}$, $\{C.4\}$, $\{A.4, B.3\}$, $\{A.4, C.4\}$, $\{B.3, C.4\}$, $\{A.4, B.3, C.4\}$
$\{A.4, B.5\}$	$\{A.4\}$, $\{B.5\}$, $\{A.4, B.5\}$
$\{B.1, C.3\}$	$\{B.1\}$, $\{C.3\}$, $\{B.1, C.3\}$

$$
\begin{aligned}
|\mathrm{RI}| &= \left(2^3-1\right)+\left(2^2-1\right)+\left(2^2-1\right)+\left(2^2-1\right)+\left(2^2-1\right) \\
&\quad +\left(2^2-1\right)+\left(2^3-1\right)+\left(2^2-1\right)+\left(2^2-1\right) \\
&= \left(2^3+2^2+2^2+2^2+2^2+2^2+2^3+2^2+2^2\right)-9 \\
&= 35
\end{aligned} \tag{5.4}
$$

5.3 RI-TREE CONSTRUCTION

As described in section 5.2.1, the maximal instances are used to generate row instances of co-locations. The RI-tree for finding maximal instances is introduced in this section.

A maximal instance m should satisfy the following two conditions: (1) m is a row instance of a co-location; (2) m cannot join with other instances to generate a high-size row instance.

An RI-tree is a kind of rooted tree. The root of an RI-tree is labeled as F (F is a set of all spatial instances). A branch of the RI-tree is constructed in a corresponding connective subgraph in the graph G. The node in the RI-tree represents the row instance of co-location patterns. The node u is the parent of the node v; when v is a row instance of size k co-locations, u is one of row instances of size k – 1 co-locations. An RI-tree is constructed to generating maximal instances, which is the key of co-location mining. A set of spatial instances with their spatial relationships (i.e., Euclidean distance) is input, and an algorithm implemented in Python outputs all maximal instances. In this process, neighbor relationships between instances of the same spatial feature type are not taken into consideration, since our goal is to find the positive relationship between different spatial feature types.

Based on the definitions of RI-tree and maximal instance, there is the following algorithm for constructing an RI-tree.

Algorithm 5.1

1. Create the root of the RI-tree, and label it as *F*.
2. Push all spatial instances into a set *F* in alphabetic and numerical descending order.
3. Pop an instance from RI_1, and delete it in RI_1, then create a child node of the root "*F*" for this instance.
4. Find out the instances that are neighbors of this instance, the different spatial features from this instance, and "bigger" than this instance.
    a. If not, return to (3).
    b. If so, push them into a set *T* in alphabetic descending order and create a child node for them.
5. Find out the instances that are neighbors of the last instance in *T*, and delete it in RI_1.
    a. If not, push *T* in *M*; Return to (3).
    b. If so and it also has neighborhood relationship with the rest of the instances in *T*, put it at the end of *T* and create a child node for them; return to (5).
    c. If so but it is not neighbors with all other instances in *T*, push *T* in *M*, create another child node of the root for the last instance in *T*, and generate a new *T* by combining it with the last instance in *T*; return to (5).
6. Repeat the operation above till RI_1 is empty.

Figure 5.2 shows the process of constructing an RI-tree for the example data set in Figure 5.1a. Leaf nodes of this RI-tree, such as $\{A.1,\ B.2,\ C.1\}$, $\{B.2,\ A.5\}$, $\{A.5,\ B.6\}$, $\{A.2,\ B.4\}$, $\{A.3,\ C.2\}$, $\{C.2,\ A.1\}$, $\{A.4,\ B.3,\ C.4\}$, $\{A.4,\ B.5\}$, and $\{B.1,\ C.3\}$ are maximal instances of the example data set. Taking $\{A.1,\ B.2,\ C.1\}$ as an example to illustrate the generation of a maximal instance, *A*.1 is a spatial instance popped from the example data set; create a child node of *A*.1 for *A*.1 and *B*.2, since there is a neighbor relationship between *A*.1 and *B*.2. Then find out the instance C.1 which has neighbor relationship with *B*.2 and check whether there is a neighbor relationship between *A*.1 and C.1. A child node is created for *A*.1, *B*.2, and C.1 since there is a neighbor relationship between *A*.1 and *C*.1. The spatial instance *D*.3 has neighbor relationship with C.1, but not with *A*.1 and *B*.2, so $\{A.1,\ B.2,\ C.1\}$ is a maximal instance.

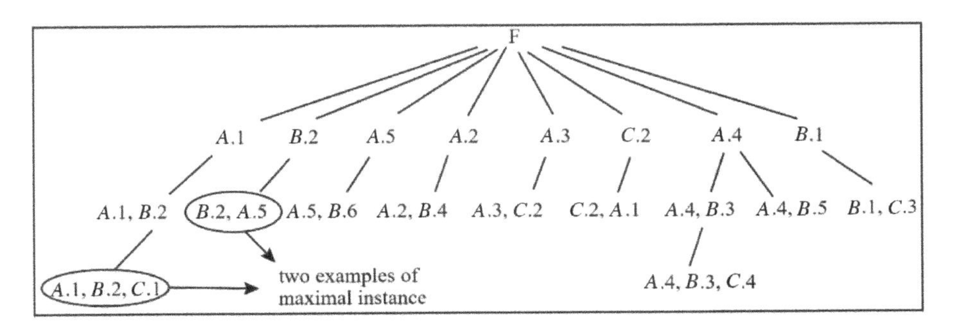

FIGURE 5.2 The process of constructing a RI-tree.

5.3.1 Rules of RI-Tree

Before building an RI-tree, all spatial instances are put into set *F*. The instance will be deleted from *F* if a child node of the root is created for it or if it has neighbor relationship with the child node of the root. For example, a child node of the root is created for *A*.5, so *A*.5 will be deleted from *F*; *B*.2 and *C*.1 are deleted from *F* since they have neighbor relationships with *A*.5 (*A*.5 is a child node of the root).

Each node is a row instance of co-location, and note that not all row instances are listed in the RI-tree, but all neighbor relationships are contained in the RI-tree. The number of row instances is very large if there are a lot of instances in a spatial data set. The aim of a RI-tree is to generate maximal instance; there is no need to list all row instances in an RI-tree. A node in RI-tree contains neighbor relationships between all instances in the node.

The nodes of the *n*th layer of a RI-tree are $n - 1$ size row instances. The spatial instances in *F* are selected as child nodes of the root, that is, nodes in the second layer of RI-tree. From layer 3 to layer *n*, there are size 2 row instances, size 3 row instances, . . . , $n - 1$ size row instances. The highest size of row instance of the example data in Figure 5.3 is 5 since the RI-tree constructed for this example data has only four layers.

After creating a child node of the root, scan its neighbor relationships. The neighbor relationships are identified by Euclidean distance metric; two instances have neighbor relationship if their Euclidean distance is less than the user-defined minimum distance threshold (Huang et al. 2004). If this node has a neighbor relationship with another instance, a child node of this node will be created under this branch for these two instances; if there is no common neighbor of instances in this node, another instance will be popped from *F* and a child node of the root will be created for it. For example, instance *B*.2 is a child node of the root; the node "*B*.2, *A*.5" is created under this branch since *B*.2 has a neighbor relationship with *A*.5. There is no common neighbor of *B*.2 and *A*.5, but *A*.5 still has a relationship with *B*.6, so a new child node of the root is created for *A*.5, and the node *A*.5 has a child node "*A*.5, *B*.6."

The RI-tree is built from left to right. When a leaf node appears in one branch of the RI-tree, another branch can be established. A node is a leaf node if there is

```
Input:
        E: a set of spatial event types
        I:    a    set    of    instances    (instance -
id, spatial feature type, location)
        min_d: a distance threshold
        min_prev: a minimal prevalent threshold
Output:
        a set of maximal instances
Variables:
        M: a set of maximal instances
        R: a spatial neighbor relation
Method:
1.      R=gen_neighborhood;
2.      push all the instances in I into a stack RI_1 in
    descending order;
3.      while (not empty RI_1) do {
4.         i ←Pop an instance from RI_1 and delete it in RI_1;
5.         for (j in RI_1) do {
6.          if ({i,j} in R) do {
7.             M.insert(0, T), T= {i,j}; }
8.             for (m in RI_1) do {
9.                if ({T[1],m} in R or {m,T[1]} in R) do {
10.                  if ({T[0],m} in R or {m,T[0]} in R) do {
11.                       M.remove(T);
12.                       T.insert(2, m);
13.                       M.insert(0, T);
14.                       RI_1.remove(m); }
15.                  }
16.               }
17.        }
18.  }
```

FIGURE 5.3 Algorithm for generating maximal instances.

no common neighbor of instances in this node, which means the end of the branch where this node is located.

The leaf nodes of a RI-tree are maximal instances, because a leaf node is a row instance in which all instances in it are neighbors and no other instances can join with it to generate a high size row instance. It is obvious that the leaf node meets the conditions of maximal instance. The number of maximal instances equals the number of branches of the RI-tree.

5.3.2 Completeness of RI-Tree

None of the neighbor relationships is omitted in the process of finding maximal instances. All the spatial instances are scanned, and all neighbor relationships between instances of different spatial types are considered in the process of constructing a RI-tree. Although not all neighbor relationships are shown in the RI-tree, they are considered when creating nodes. For example, the node "*A*.1, *B*.2, *C*.1" contains the neighbor relationships between *A*.1 and *B*.2, *A*.1 and *C*.1, and *B*.2 and *C*.1, although the neighbor relationships between *A*.1 and *C*.1, *B*.2 and *C*.1 are not shown in the RI-tree, and the neighbor relationships are not duplicated. Ince all neighbor relationships of an instance are identified, this instance will be deleted in *F*.

5.4 GENERATION OF CO-LOCATIONS

Candidate co-location generation: The basic algorithms rely on a combinatorial approach and use apriori_gen (Agrawal and Srikant 1996) to generate size k + 1 candidate co-locations from size k prevalent co-locations. However, this chapter generates candidate co-locations from maximal instances, which does not require any join operations. $\mathrm{C} = \{e_1, e_2, \cdots, e_k\}$ is a candidate co-location, if $\mathrm{m} = \{i_1, i_2, \cdots, i_k\}$ is maximal instance, i_n is the instance of spatial event types e_n, $1 \le n \le k$. For example, {*A*.1, *B*.2, *C*.1} is a maximal instance, so $\mathrm{T} = \{A, B, C\}$ is a candidate co-location. And candidate co-locations are also generated from candidate co-locations that are not prevalent. Those candidate co-locations generated from maximal instances are not all prevalent. Candidate co-locations which fail to be a prevalent pattern are selected; their subsets are corresponding low-size candidate co-locations. That is because of the anti-monotonicity of participation index. If $pi(C) < min\_prev$ (C is a candidate co-location), there must exist $pi(c) \ge pi(C)$ (c is a subset of C), c may be a prevalent co-location, so c is a candidate co-location. Taking Figure 5.1 as an example, if $\mathrm{min\_prev}$ is set to 0.4, $\mathrm{T} = \{A, B, C\}$ is not a prevalent co-location since $pi(T) = min\{pr(T, A), pr(T, B), pr(T, C)\} = min(2/5, 2/6, 2/4) = 1/3 < 0.4$. Therefore, {*A*, *B*}, {*A*, *C*}, and {*B*, *C*} (subsets of {*A*, *B*, *C*}) are candidate co-locations.

Pruning: In order to make our algorithm more efficient, this step introduces the pruning strategies, which can greatly reduce the unnecessary calculate time. Prevalence-based pruning (2004) is also used in our algorithm. The refinement filtering of co-locations is done by the participation index values calculated from their co-location instances. Prevalent co-locations satisfying a given threshold are selected. Actually, there is no need to calculate the participation index of each candidate co-location. Because the participation ratio and participation index are antimonotone (monotonically nonincreasing) as the size of the co-location increases (Huang et al. 2004), if a high-size candidate co-location is prevalent, subsets of this candidate co-location are all prevalent. Therefore, participation indexes of high-size candidate co-locations (generated by maximal instances) are prioritized. If the participation index values are above a given threshold, there is no need to calculate the participation indexes of subsets of high-size candidate co-location; if the participation index values are below a given threshold, participation indexes of subsets of high-size candidate co-locations need to be calculated.

The second pruning strategy can be explained by *Lemma 4*.

Lemma 4

Subsets of a candidate co-location are prevalent if it is a prevalent co-location.
Proof. This lemma is proved by the antimonotone property of participation index. C is a candidate co-location, C' is a subset of C, k and n ($k > n$) are the number of instances in C and the number of instances in C'. From the previous formula, it can be concluded that the participation index of a candidate co-location is less than or equal to that of its subset. If $pi(C) < min\_prev$, then $pi(C') < min\_prev$ must be established, so the conclusion can be drawn: subsets of a candidate co-location are prevalent if it is a prevalent co-location.

$$\begin{aligned} pi(C) = min\left\{\frac{\left|\pi_{e_1}\left(table\_instance(C)\right)\right|}{|e_1|},\ldots,\frac{\left|\pi_{e_k}\left(table\_instance(C)\right)\right|}{|e_k|}\right\} \\ = min_{i=1}^{k}\left\{pr(C,e_i)\right\} \\ \le min_{i=1}^{n}\left\{pr(C,e_i)\right\} \\ \le min_{i=1}^{n}\left\{pr(C',e_i)\right\} = pi(C') \end{aligned} \tag{5.5}$$

Quite a lot of maximal instances are generated in the process of mining co-locations from the real data set introduced in section 6.5.1. $\{A, I, T, P, R\}$ is a candidate co-location since there are row instances of $\{A, I, T, P, R\}$ in maximal instances, and $\{A, I, T, P, R\}$ is the highest size candidate co-location since there are only five facility types in the real data set. The participation index of $\{A, I, T, P, R\}$ is calculated first. When min_d is set to 1, $pi(\{A, I, T, P, R\}) = 0.1$. $\{A, I, T, P, R\}$ failed to be a prevalent co-location if min_prev is set to 0.4. Therefore, $\{A, I, T, P\}$, $\{A, I, T, R\}$, $\{A, I, P, R\}$, $\{A, T, P, R\}$, and $\{I, T, P, R\}$, which are the subsets of $\{A, I, T, P, R\}$, become candidate co-locations. In the subsets of $\{A, I, T, P, R\}$, those row instances that contain 4 instances are row instances of size 4 co-locations. And the size 4 maximal instances are also row instances of size 4 co-locations. By calculating their participation indexes, $\{A, T, P, R\}$ is a prevalent co-location with $pi(\{A, T, P, R\}) = 0.48$. According to the anti-monotonicity of participation index, the participation indexes of $\{A, T, P\}$, $\{A, T, R\}$, $\{A, P, R\}$, and $\{T, P, R\}$ must be greater than 0.48. $\{A, T, P\}$, $\{A, T, R\}$, $\{A, P, R\}$, and $\{T, P, R\}$ are prevalent co-locations, so there is no need to calculate their participation indexes, which saves a lot of time. This is the pruning strategy mentioned earlier.

5.5 DISCUSSIONS FOR MAXIMAL INSTANCE ALGORITHMS

The idea of the presented algorithm is different from the existing algorithms. Their differences are:

- The existing methods identify high-size row instances from low-size row instances, but the presented algorithm first identifies the highest-size row instances (maximal instances) based on the relationship of spatial instances, then generates all low-size row instances from highest-size row instances.

- The existing methods generate candidate patterns by joining row instances, but the presented algorithm generates candidate patterns from maximal instances without join operations.

5.5.1 Comparison Analysis of Row Instance Generation

The reason why our method for generating all row instances is more efficient than join-based algorithm is demonstrated in detail here. Let $T_{gen\_row\ instance}$ and $T'_{gen\_row\ instance}$ be the computation cost of generating all row instances in join-based algorithms and our method. The following equations show the computation cost functions:

$$T_{gen\_row\ instance} = T_{scan(neighobr\ relationship)} + (k-1) \times T_{join} \tag{5.6}$$

$$T'_{gen\_row\ instance} = T_{gen\_m} + T_{gen\_instance} \tag{5.7}$$

T_{join} represents the computation cost of a join operation, T_{gen_m} represents the computation cost of generate all maximal instances, $T_{gen\_instance}$ represents the computation cost of finding subsets of maximal instances. $T_{gen\_m} \approx T_{scan(neighobr\ relationship)}$ because of generating maximal instances is also a scanning process of neighbor relationships, and $T_{gen\_instance} \ll (k-1) \times T_{join}$.

Therefore,

$$\begin{aligned}\frac{T_{gen\_row\ instance}}{T'_{gen\_row\ instance}} &= \frac{T_{scan(neighobr\ relationship)} + (k-1) \times T_{join}}{T_{gen\_m} + T_{gen\_instance}} \\ &\approx \frac{T_{scan(neighobr\ relationship)} + (k-1) \times T_{join}}{T_{scan(neighobr\ relationship)} + T_{gen\_instance}} \\ &> 1\end{aligned} \tag{5.8}$$

Apparently, $T_{gen\_row\ instance} > T'_{gen\_row\ instance}$. And a conclusion can be drawn: our method for generating all row instances is more efficient than join-based algorithm.

5.5.2 Comparison Analysis of Maximal Instance Algorithms

The computational cost of our algorithm with three basic algorithms is compared here. Each algorithm has an initial step of materializing neighborhoods. The join-based algorithm first gathers the neighbor pairs per candidate co-location, the partial join algorithm first generates the disjoint clique neighborhoods using a simple grid partitioning, the join-less algorithm first generates the star neighborhoods, and our algorithm first generates maximal instances. S represents a spatial data set, $T_{neighbor\_pairs}(S)$, $T_{clique\_neighborhoods\&cut\_relations}(S)$, $T_{star\_neighborhoods}(S)$ and $T_{gen\_m}(S)$ represent the computational costs of initial steps in each algorithm, and $T_{jb}(2)$, $T_{pj}(2)$ and $T_{jl}(2)$ represent the cost for generating size 2 co-locations in join-less algorithm, the partial join algorithm, and the join-based algorithm, respectively. Let T, T_{jl}, T_{pj} and T_{jb} be the total computation cost of our algorithm, the join-less algorithm, the partial join algorithm, and the join-based algorithm, respectively. The following equations show the total cost functions:

$$T = T_{gen\_m}(S) + T_{gen\_instance}$$
$$T_{jl} = T_{star\_neighborhoods}(S) + T_{jl}(2) + \sum\nolimits_{k>2} T_{jl}(k)$$
$$T_{pj} = T_{clique\_neighborhoods\&cut\_relations}(S) + T_{pj}(2) + \sum\nolimits_{k>2} T_{pj}(k)$$
$$T_{jb} = T_{neighbor\_pairs}(S) + T_{jb}(2) + \sum\nolimits_{k>2} T_{jb}(k) \tag{5.9}$$

Here, k is greater than 2.

Yoo and Shekhar (Yoo et al. 2004) proposed the following comparative relationships:

$$\begin{aligned} & T_{clique\_neighborhoods\&cut\_relations}(S) + T_{pj}(2) \\ & > T_{star\_neighborhoods}(S) + T_{jl}(2) \\ & > T_{neighbor\_pairs}(S) + T_{jb}(2) \end{aligned} \tag{5.10}$$

Because of: (1) the join-less algorithm has an additional cost to materialize the star neighborhoods from the neighbor pairs compared with a join-based algorithm; (2) an additional cost is required to find all size 2 inter-instances with cut relationships in a partial join algorithm; the overall cost is expected to be a little bigger than the cost to generate the star neighborhood set.

Lemma 5

$$\boldsymbol{T}_{star\_neighborhoods}(\boldsymbol{S}) > \boldsymbol{T}_{gen\_m}(S).$$

Proof. The star neighborhood partition model in a join-less algorithm takes each spatial object as a central object and finds out the remaining objects in their neighborhoods. respectively. Thus, this process takes n times; n is the number of spatial objects. The process of generating all maximal instances needs to be executed

$|C_m|$ times, $|C_m|$ is the number of maximal instances. Apparently, $n > |C_m|$, so the star neighborhood partition model in join-less method is more time consuming than the process of generating maximal instance in our method.

$$\begin{aligned} \frac{T_{star\_neighborhoods}(S)}{T_{gen\_m}(S)} & \approx \frac{n \times t_{scan(neighbor\ relationship)}}{|C_m| \times t_{scan(neighbor\ relationship)}} \\ & = \frac{n}{|C_m|} \\ & > 1 \end{aligned} \tag{5.11}$$

Lemma 6

$$\boldsymbol{T}_{gen\_m}(\boldsymbol{S}) > \boldsymbol{T}_{neighbor\_pairs}(\boldsymbol{S}) + \boldsymbol{T}_{jb}(2).$$

Proof. Gathering the neighbor pairs per candidate co-location is the most basic operation in co-location mining. The neighbor pairs are used to generate maximal instances; it has an additional cost compared with gathering the neighbor pairs per candidate co-location and calculating their prevalence measures.

According to *Lemma 5* and *Lemma 6*, the following equation can be reached:

$$\begin{aligned} & T_{clique\_neighborhoods\&cut\_relations}(S) + T_{pj}(2) \\ & > T_{star\_neighborhoods}(S) + T_{jl}(2) \\ & > T_{gen\_m}(S) \\ & > T_{neighbor\_pairs}(S) + T_{jb}(2) \end{aligned} \tag{5.12}$$

The advantage of maximal instance algorithm lies in the latter part. Three basic algorithms generate bigger size co-locations size by size after generating size 2 co-locations. This operation needs to repeat $k - 2$ times to generate all co-locations. Each operation cannot be omitted. And cross-size operation is not allowed (size k co-locations cannot be directly generated from size $k - 2$ co-locations' size k co-locations are generated from size $k - 1$ co-locations after generating size $k - 1$ co-locations from size $k - 2$ co-locations). Therefore, the execution time of each join operation should be added to the total computational costs of three basic algorithms.

However, once maximal instance algorithm generates maximal instances, row instances of all size co-locations can be generated from subsets of maximal instances. There is no need for the join operation. A large amount of time consumption is avoided in this part; the bigger the size of co-locations, the more obvious the effect is. Therefore, there are the following equations:

$$T_{gen\_instance} \ll \sum\nolimits_{k>2} T_{jl}(k), \tag{5.13}$$

$$T_{gen\_instance} \ll \sum\nolimits_{k>2} T_{pj}(k), \tag{5.14}$$

$$T_{gen\_instance} \ll \sum\nolimits_{k>2} T_{jb}(k). \tag{5.15}$$

On the basis of the equations 5.13, 5.14, 5.15, and 5.16, we can say that our algorithm has a lower computational cost than three basic algorithms.

$$T < T_{jl}, \tag{5.16}$$

$$T < T_{pj}, \tag{5.17}$$

$$T < T_{jb}. \tag{5.18}$$

REFERENCES

Adilmagambetov, A., Zaiane, O.R., and Osornio-Vargas, A., Discovering co-location patterns in datasets with extended spatial objects. *International Conference on Data Warehousing & Knowledge Discovery*, Springer Berlin Heidelberg, Prague, Czech Republic, August 26–29, 2013, pp. 84–96.

Agrawal, R., and Srikant, R., *Fast algorithms for mining association rules: Readings in database systems* (3rd ed.). Morgan Kaufmann Publishers Inc., 1996. ISBN:978–1558605237.

Albanese, A., Pal, S.K., and Petrosino. A., A rough set approach to spatio-temporal outlier detection. *Fuzzy Logic and Applications*. Springer Berlin, Heidelberg, 2011. ISBN: 978-3-642-23712-6.

Alnaymat, G., Chawla, S., and Schickler, W., Enumeration of maximal clique for mining spatial co-location patterns. *IEEE/ACS International Conference on Computer Systems & Applications IEEE*, Doha, Qatar, March 31–April 4, 2008, pp. 126–133.

Bao, X., and Wang, L., Discovering interesting co-location patterns interactively using ontologies. *International Conference on Database Systems for Advanced Applications*, Suzhou, China, March 22, 2017, pp. 75–89.

Celik, M., Azginoglu, N., and Terzi, R., Mining periodic spatio-temporal co-occurrence patterns: A summary of results. *Innovations in Intelligent Systems and Applications (INISTA), 2012 International Symposium on IEEE*, Trabzon, Turkey, July 2–4, 2012, pp. 1–5.

Celik, M., Kang, J.M., and Shekhar. S., Zonal co-location pattern discovery with dynamic parameters. *Data Mining, 2007. ICDM 2007. Seventh IEEE International Conference on IEEE*, Omaha, NE, USA, October 28–31, 2007, pp. 433–438.

Celtic, M., Discovering partial spatio-temporal co-occurrence patterns. *IEEE International Conference on Spatial Data Mining and Geographical Knowledge Services, ICSDM 2011*, Fuzhou, China, IEEE, June 29–July 1, 2011.

Deng, M., Cai, J., Liu, Q., He, Z., and Tang, J., Multi-level method for discovery of regional co-location patterns. *International Journal of Geographical Information Science*, 2017, pp. 1–25.

Ding, W., Jiamthapthaksin, R., Parmar, R., Jiang, D., Stepinski, T.F., and Eick, C.F., Towards region discovery in spatial datasets. *Pacific-Asia Conference on Advances in Knowledge Discovery & Data Mining*, Springer-Verlag, Osaka, Japan, May 20–23, 2008, pp. 88–99.

Huang, Y., Pei, J., and Xiong, H., Mining co-location patterns with rare events from spatial data sets. *GeoInformatica*, vol. 10, no. 3, 2006, pp. 239–260.

Huang, Y., Shekhar, S., and Xiong, H., Discovering colocation patterns from spatial data sets: A general approach. *IEEE Transactions on Knowledge and Data Engineering*, vol. 16, no. 12, 2004, pp. 1472–1485.

Huang, Y., Xiong, H., Shekhar, S., and Pei, J., Mining confident co-location rules without a support threshold. *ACM Symposium on Applied Computing, 2003*. Melbourne, Florida, USA, March 9, 2003, pp. 497–501.

Huang, Y., Zhang, P., and Zhang, C., On the relationships between clustering and spatial co-location pattern mining. *International Journal on Artificial Intelligence Tools*, vol. 17, no. 1, 2008, pp. 55–70.

Jiang, Y., Wang, L., and Chen, H., Discovering both positive and negative co-location rules from spatial data sets. *Software Engineering and Data Mining (SEDM), 2010 2nd International Conference on IEEE*, Chengdu, China, June 23–25, 2010, pp. 398–403.

Kim, S.K., Kim, Y., and Kim, U., Maximal cliques generating algorithm for spatial colocation pattern mining. *FTRA International Conference on Secure and Trust Computing, Data Management, and Application 2011*, Loutraki, Greece, June 28–30, 2011, pp. 241–250.

Kim, S.K., Lee, J.H., Ryu, K.H., and Kim. U., A framework of spatial co-location pattern mining for ubiquitous GIS. *Multimedia Tools and Applications*, vol. 71, no. 1, 2014, pp. 199–218.

Lei, P.R., Li, S.C., and Peng, W.C., QS–STT: QuadSection clustering and spatial–temporal trajectory model for location prediction. *Distributed and Parallel Databases (2013)*, vol. 31, no. 2, October 20, 2012, pp. 231–258.

Li, F., Cheng, D., Hadjieleftheriou, M., Kollions, G., and Teng, S.H., On trip planning queries in spatial databases. *Advances in Spatial and Temporal Databases. SSTD 2005. Lecture Notes in Computer Science*, vol. 31, no. 1, 2004, pp. 1–16.

Li, Q., Research on algorithm of positive/negative co-location pattern mining in spatial data. *MS. Thesis*, Guilin University of Technology, July 2020.

Mohan, P., Shekhar, S., Shine, J.A., Rogers, J.P., and Wayant, N., A neighborhood graph based approach to regional co-location pattern discovery: A summary of results. *19th ACM SIGSPATIAL International Symposium on Advances in Geographic Information Systems*, ACM–GIS 2011, Chicago, IL, USA, Proceedings ACM, November 1–4, 2011.

Ouyang, Z., Wang, L., and Wu. P., Spatial co-location pattern discovery from fuzzy objects. *International Journal on Artificial Intelligence Tools*, vol. 26, no. 2, 2017, 1750003, pp. 1–21.

Qian, F., He, Q., and He, J., Mining spatial co-location patterns with dynamic neighborhood constraint. *Machine Learning and Knowledge Discovery in Databases, European Conference, ECML PKDD 2009*, Bled, Slovenia, Proceedings, Part II DBLP, September 7–11, 2009.

Shekhar, S., and Huang, Y., Discovering spatial colocation patterns: A summary of results. *Advances in Spatial and Temporal Databases, 7th International Symposium, SSTD 2001*, Redondo Beach, CA, USA, July 12–15, Proceedings DBLP, 2001.

Sundaram, V.M., Thnagavelu, A., and Paneer, P., Discovering co-location patterns from spatial domain using a Delaunay approach. *Procedia Engineering*, vol. 38, 2012, pp. 2832–2845. ISSN:1877–7058.

Venkatesan, M., and Thangavelu, A., *A multiple window-based co-location pattern mining approach for various types of spatial data.* Inderscience Publishers, Geneva, Swizterland, 2013. ISBN:0952–8091.

Verhein, F., and Al-Naymat, G., Fast mining of complex spatial co-location patterns using GLIMIT. *Data Mining Workshops, ICDM Workshops 2007, Seventh IEEE International Conference on*, Omaha, NE, USA, October 28–31, 2007, pp. 679–684.

Wang, L.Z., Bao, X., Chen, H., and Cao, L., Effective lossless condensed representation and discovery of spatial co-location patterns. *Information Sciences (2018)*, S0020025518300148, Vol. 435–437, 2018, ISSN 0020–0255, pp. 197–213.

Williams, J., Shore, S.E., and Foy. J.M., Co-location of mental health professionals in primary care settings: Three North Carolina models. *Clinical Pediatrics*, vol. 45, no. 6, 2006, pp. 537–543.

Xiao, X., Efficient co-location pattern discovery. *Hong Kong University of Science and Technology (People's Republic of China)*, 2009.

Yang, Z.F., and Tang, H.W., SNN neighbor and SNN density-based co-location pattern discovery. *International Conference on E-business & E-government*, Shanghai China, May 6–8, 2011, pp. 1–5.

Yao, X., Peng, L., Yang, L., and Chi, T., A fast space-saving algorithm for maximal co-location pattern mining. *Expert Systems with Applications* (2016): S0957417416303517, vol. 63, 2016, SSN 0957–4174, pp. 310–323.

Yao, X., Wang, D., Peng, L., and Chi, T., An adaptive maximal co-location mining algorithm. *IGARSS 2017–2017 IEEE International Geoscience and Remote Sensing Symposium IEEE*, Fort Worth, TX, USA, July 23–28, 2017, pp. 1646–1649.

Yoo, J.S., Jin, D., Boulware, D., and Kimmey, D., A parallel spatial co-location mining algorithm based on MapReduce. *2014 IEEE International Congress on Big Data (BigData Congress) IEEE*, Anchorage, AK, USA, June 27–July 2, 2014, pp. 25–31.

Yoo, J.S., and Shekhar, S., A partial join approach for mining co-location patterns: A summary of results. *Association for Computing Machinery*, New York, USA GIS'04, November 12–13, 2004, Copyright 2004 ACM 1-58113-979-9/04/0011, pp. 241–249.

Yoo, J.S., and Shekhar, S.A., Joinless approach for mining spatial colocation patterns. *IEEE Transactions on Knowledge and Data Engineering*, vol. 18, no. 10, 2006, pp. 1323–1337.

Yoo, J.S., Shekhar, S., and Celik, M., A join-less approach for co-location pattern mining: A summary of results. *Data Mining, Fifth IEEE International Conference on IEEE*, Houston, TX, USA, November 27–30, 2005, pp. 813–816.

Yu, P., FP-tree based spatial co-location pattern mining. *Diss.* University of North Texas, Denton, Texas, 2005.

Yu, W., Identifying and analyzing the prevalent regions of a co-location pattern using polygons clustering approach. *ISPRS International Journal of Geo-Information*, vol. 6, no. 9, 2017, pp. 1–22.

Zhang, T., Association rules. *Pacific-Asia Conference on Knowledge Discovery & Data Mining Springer*, Berlin, Heidelberg, 2000. ISBN: 978-3-540-67382-8.

Zhou, G., Chen, W., and Kelmelis, J., A comprehensive study on urban true orthorectification. *IEEE Transactions on Geoscience and Remote Sensing*, vol. 43, no. 9, 2005, pp. 2138–2147.

Zhou, G., and Wang, L., Co-location decision tree for enhancing decision-making of pavement maintenance and rehabilitation. *Transportation Research Part C: Emerging Technologies*, vol. 21, no. 1, 2012, pp. 287–305.

Zhou, G., and Wang, L., GIS and data mining to enhance pavement rehabilitation decision-making. *Journal of Transportation Engineering*, vol. 136, no. 4, February 2010, pp. 332–341.

6 Negative Co-Location Pattern Mining Algorithms

6.1 INTRODUCTION

The co-location pattern, as a method for mining spatial data, has received increasing attention since Yoo and Shekhar first proposed two co-location algorithms based on join and joinless modes (Morimoto 2001; Shekhar and Huang 2001; Huang et al. 2004; Yoo et al. 2004; Yoo et al. 2005). Many scholars have afterward presented new co-location algorithms, for instance, fuzzy co-location pattern mining, parallel co-location pattern mining, the adaptive maximal co-location algorithm, efficient co-location pattern mining, co-location pattern mining with rare features (Bao and Wang 2019; Wang et al. 2008, 2009; He et al. 2015), the co-location-based decision tree (CL-DT) (Zhou et al. 2012, 2014, 2016), manifold learning co-location pattern mining (Zhou et al. 2016, 2018), and maximal instance co-location pattern mining (Zhou 2011; Zhou et al. 2021).

Although the advance of mining co-location patterns has been made, there are still some potentially useful patterns that have not been fully mined, such as negative co-location patterns. Spatial data mining with negative co-location patterns can be significant because it can find features with strong negative correlations and determine mutually exclusive relationships between spatial features, which can play a vital role in many applications. For example, lilacs planted next to cloves wither immediately. The scent of cloves can also endanger the narcissus. If you keep lilacs, violets, tulips, and forget-me-nots together, each suffers. Therefore, garden designs and forest plans need to fully consider the relationships among plants to reap benefits and avoid harm. Similarly, negative co-location patterns can play great roles in biology. They can identify a variety of cells that inhibit each other and can be applied in vaccines and pharmaceuticals.

A few scholars have proposed mining algorithms for association rules and sequences for negative co-location. For example, Wu et al. (2004) proposed efficient mining of both positive and negative association rules. Zheng et al. (2009) proposed some constraint conditions and mining algorithms for negative sequence patterns. Cao et al. (2016) proposed the e-NSP (efficient negative sequential patterns) algorithm, which can effectively identify negative sequence patterns. Dong et al. (2018) proposed the f-NSP (fast negative sequential patterns) algorithm.

This chapter first overviews the existing algorithms and methods, which are especially presented by Yang et al. (2010), and then proposes a new algorithm for mining

DOI: 10.1201/9781003139416-6

negative co-location patterns. The proposed method is based on joining low prevalent co-location patterns or candidate negative co-location patterns to generate all candidate negative co-location patterns, and a paradigm is detailed for illustrating the algorithm proposed in this chapter.

6.2 DEFINITION AND LEMMA FOR NEGATIVE CO-LOCATION

6.2.1 Basic Definition of Negative Co-Location

Negative co-location mining is to find out the subsets of spatial features that are strong negative associated. For the spatial data, each spatial instance is recorded as $T.i$, where T is the spatial feature type of the spatial instance and i is the unique ID of the instance within each spatial feature type. In the negative co-location pattern mining, the Euclidean distance is still used to measure the proximity between spatial instances, that is, $\mathrm{R}(A.1, B.1) \leftrightarrow (distance(A.1, B.1) \le d)$. In Figure 6.1, the real lines are used to connect the neighboring instances, such as $A.1$ and $B.2$.

The basic concepts pertaining to negative co-location pattern mining that follow were given by Jiang et al. (2010).

Definition 1

Negative co-location patterns: A co-location pattern T is defined as a negative co-location pattern if $T = X \cup \overline{Y}$, where X is a set of positive items (positive spatial features), $\overline{Y}$ is a set of negative items (negative spatial features), and $|\overline{Y}| \ge 1$, $X \cap Y = \varnothing$ (Jiang et al. 2010).

Definition 2

Participation index (PI) of negative co-location patterns: $T = X \cup \overline{Y}$ is defined as $PI(T) = min_{i=1}^{k}\{PR(T, X_i)\}$, where $PR(T, X_i)$ is the participation ratio (PR) of spatial feature X_i in a negative co-location pattern T (Jiang et al. 2010).

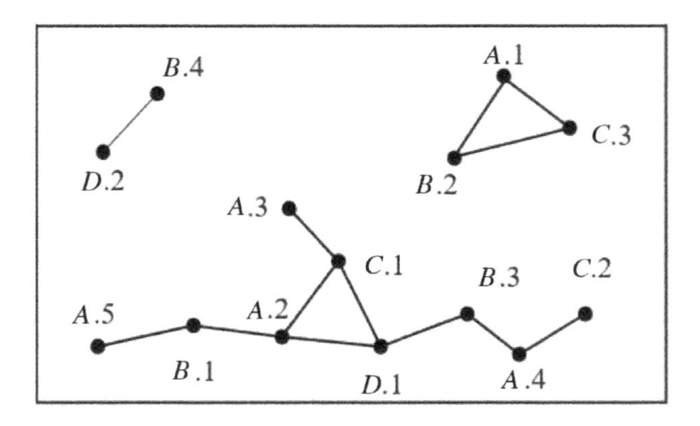

FIGURE 6.1 An example of a spatial data set.

Definition 3

Participation ratio (PR): The participation ratio, noted as $PR(T, X_i)$, of spatial feature X_i in a negative co-location $T = X \cup \overline{Y}$ is defined as and can be computed by (Jiang et al. 2010):

$$PR(T, X_i) = \frac{\left|\pi_{X_i}\left(table\_instance(T)\right)\right|}{\left|table\_instance(X_i)\right|} \tag{6.1}$$

Definition 4

Prevalent negative co-location patterns: For a given minimum prevalent threshold (*min_prev*), a negative co-location pattern $T = X \cup \overline{Y}$ is defined as a prevalent negative co-location pattern if T meets the following conditions (Jiang et al. 2010).

1. $PI(X) \geq \min\_prev, PI(Y) \geq \min\_prev \text{ and } PI(X \cup Y) < \min\_prev$
2. $PI(T) \geq \min\_prev$

6.2.2 Lemmas for Negative Co-Location

Jiang et al. (2010) also presented the following lemmas for negative co-location mining.

Lemma 1

Let $T = X \cup \overline{Y}$ be a candidate prevalence negative co-location pattern, where $X = \{X_1, X_2, \ldots, X_k\}$. If $A_1, A_2, \ldots, A_k$ are respectively the number of instances of spatial features $X_1, X_2, \ldots, X_k$ in the positive co-location $X \cup Y$, the PI of negative co-location $X \cup \overline{Y}$ is (Jiang et al. 2010):

$$PI(X \cup \overline{Y}) = min\left\{\frac{|X_1| - A_1}{|X_1|}, \frac{|X_2| - A_2}{|X_2|}, \ldots, \frac{|X_k| - A_k}{|X_k|}\right\} \tag{6.2}$$

where $|X_i|$ is the number of instances of spatial feature X_i.

Lemma 2

Let $X' \cup \overline{Y}$ and $X'' \cup \overline{Y}$ be prevalent negative co-locations. For prevalent positive co-location pattern $X = X' \cup X''$, $X \cup \overline{Y}$ must be a prevalent negative co-location pattern (Jiang et al. 2010).

Lemma 3

Let $X \cup \overline{Y}$ be a prevalent negative co-location pattern. For prevalent positive co-location pattern Z, if $Y \subseteq Z$ holds, then $X \cup \overline{Z}$ is also a prevalent negative co-location pattern (Jiang et al. 2010).

6.3 ALGORITHM FOR MINING NEGATIVE CO-LOCATION PATTERNS

In order to find out the potentially useful negative co-location patterns in spatial data sets, Jiang et al. (2010) presented a few new concepts and lemmas as follows.

Definition 5

Spatial relation pair: If there is a size 2 prevalent co-location pattern or a size 2 prevalent negative co-location pattern, this is defined as a neighbor relation pair (Jiang et al. 2010).

Lemma 4

The participation ratio (PR) and the participation index (PI) are monotonically nondecreasing as the size of negative co-location increases (Jiang et al. 2010).

Proof:

1. Supposing that $M = X \cup \overline{Y}$, $N = X' \cup \overline{Y}$, $X' \subseteq X$, A_i represents the number of instances of each feature X_i in $X \cup \overline{Y}$, B_i represents the number of instances of each feature X_i in $X' \cup \overline{Y}$, $A_i \le B_i$ since the spatial feature instance that participates in a row instance of X also participates in a row instance of X'. Therefore, $PR(M, X_i) = \frac{|X_i| - A_i}{|X_i|} \ge \frac{|X_i| - B_i}{|X_i|} = PR(N, X_i)$ and $PI(M) = min\left(\frac{|X_i| - A_i}{|X_i|}\right) \ge min\left(\frac{|X_i| - B_i}{|X_i|}\right) = PI(N)$.
2. Supposing that $M = X \cup \overline{Y}$, $N = X \cup \overline{Y'}$, $Y' \subseteq Y$, A_i represents the number of instances of each feature X_i in $X \cup \overline{Y}$, B_i represents the number of instances of each feature X_i in $X' \cup \overline{Y}$, $A_i = B_i$ since the spatial feature instance that participates in a row instance of X also participates in a row instance of X'. Therefore, $PR(M, X_i) = \frac{|X_i| - A_i}{|X_i|} = \frac{|X_i| - B_i}{|X_i|} = PR(N, X_i)$ and $PI(M) = min\left(\frac{|X_i| - A_i}{|X_i|}\right) = min\left(\frac{|X_i| - B_i}{|X_i|}\right) = PI(N)$.

According to (1) and (2), the participation ratio and the participation index are monotonically nondecreasing as the size of negative co-location increases.

Lemma 5

Let $X \cup \overline{Y}$ be a prevalent negative co-location; if $PI(X') \ge min\_prev$ and $PI(Y') \ge min\_prev$ holds, then $X' \cup \overline{Y}$ and $X \cup \overline{Y'}$ are also prevalent co-locations, where $X \subseteq X'$ and $Y \subseteq Y'$ (Jiang et al. 2010).

Proof:

$PI(X) \ge min\_prev$, $PI(Y) \ge min\_prev$, $PI(X \cup \overline{Y}) \ge min\_prev$. and $PI(X \cup Y) < min\_prev$, since $X \cup \overline{Y}$ is a prevalent negative co-location. $PI(X' \cup Y) < PI(X \cup Y) <$

min_prev and $PI(X \cup Y') < PI(X \cup Y) < min\_prev$ since the participation index is antimonotone (monotonically nonincreasing) as the size of the positive co-location increases. It can be concluded from ***Lemma 4*** that $PI(X' \cup \overline{Y}) \geq PI(X \cup \overline{Y}) \geq min\_prev$ and $PI(X \cup \overline{Y'}) \geq PI(X \cup \overline{Y}) \geq min\_prev$. Therefore, $X' \cup \overline{Y}$ and $X \cup \overline{Y'}$ are also prevalent co-locations, where $X \subseteq X'$ and $Y \subseteq Y'$.

6.3.1 Generation of Candidate Negative Co-Locations

Jiang et al. (2010) used non-prevalent co-locations to generate candidate negative co-locations. This method is very time consuming, because first of all, those non-prevalent co-locations should be found from the process of co-location pattern mining, and then the negative correlation items of these non-prevalent co-location patterns are combined to generate candidate negative co-location patterns. For example, $\{A,B,C\}$ is a non-prevalent co-location pattern, so

$\{\overline{A},B,C\},\{A,\overline{B},C\},\{A,B,\overline{C}\},\{\overline{A},\overline{B},C\},\{\overline{A},B,\overline{C}\}, \ldots$ are the candidate negative co-location patterns. By calculating their participation indexes, it can be determined whether they are frequent negative co-location patterns. A non-prevalent co-location pattern can form multiple candidate negative co-location patterns, and it takes too much time to calculate their participation indexes one by one.

In order to generate candidate negative co-location patterns quickly and effectively and avoid the time-consuming problem caused by unnecessary calculation of the participation index, this chapter proposes a new method to generate candidate negative co-location patterns, that is, generate candidate negative co-locations by joining size 2 prevalent co-locations and size 2 prevalent negative co-locations. Size 2 prevalent co-locations and size 2 prevalent negative co-locations are called a spatial relation pair. Because this proposed algorithm takes little time to generate the size 2 prevalent co-locations and size 2 prevalent negative co-locations according to the neighbor relationships between spatial instances, it is very effective to generate candidate negative co-location patterns by using them. For example, $\{A,\overline{C}\}$ and $\{B,\overline{C}\}$ are size 2 prevalent co-locations. The $\{A,B,\overline{C}\}$ obtained by joining $\{A,\overline{C}\}$ and $\{B,\overline{C}\}$ is a candidate negative co-location pattern (Figure 6.2). This method can mine the negative co-location pattern without generating the co-location pattern, which reduces a lot of computing time.

Take the example data in Figure 6.1 as an example to further explain the generation of candidate negative co-locations. First, the participation indexes of the size 2 co-locations are calculated according to the proximity of each instance. As shown in Table 6.1, the participation indexes of co-location patterns $\{A,B\}$, $\{A,C\}$, $\{A,D\}$, $\{B,C\}$, $\{B,D\}$, and $\{C,D\}$ are 3/4, 4/5, 1/5, 1/4, 1/2, and 1/3, respectively. If the minimum frequency threshold min_prev is set to 0.6, then co-locations $\{A,B\}$ and $\{A,C\}$ are prevalent. The non-prevalent co-locations $\{A,D\}$, $\{B,C\}$, $\{B,D\}$, and $\{C,D\}$ provide the possibility for the generation of candidate negative co-location patterns. For example, $\{A,D\}$ is non-prevalent, so $\{A,\overline{D}\}$ and $\{\overline{A},D\}$ are the candidate negative co-location patterns. According to the calculation method of the participation index of negative co-locations, the participation index of each candidate

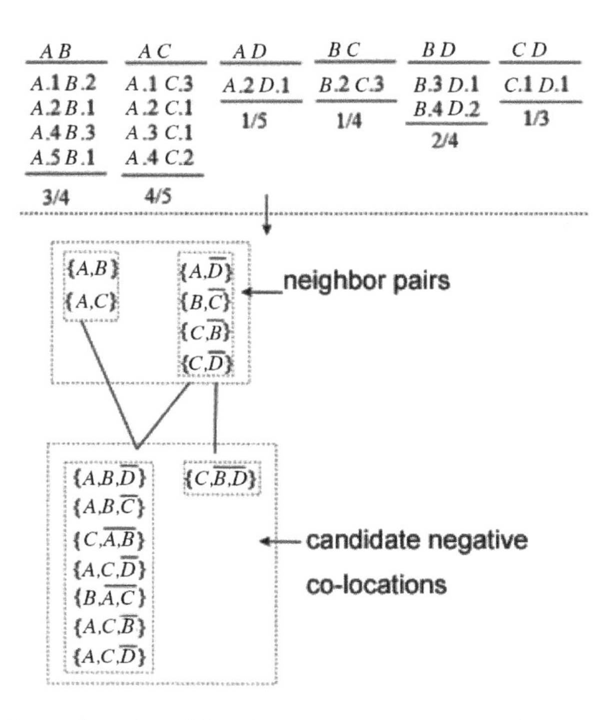

FIGURE 6.2 Generation of candidate size 2 prevalent negative co-locations.

TABLE 6.1
Table Instances and Participation Indexes of Size 2 Co-Locations

Co-location pattern	$\{A,B\}$	$\{A,C\}$	$\{A,D\}$	$\{B,C\}$	$\{B,D\}$	$\{C,D\}$
Table instance	$\{A.1,B.2\}$	$\{A.1,C.3\}$	$\{A.2,D.1\}$	$\{B.2,C.3\}$	$\{B.3,D.1\}$	$\{C.1,D.1\}$
	$\{A.2,B.1\}$	$\{A.2,C.1\}$			$\{B.4,D.2\}$	
	$\{A.4,B.3\}$	$\{A.3,C.1\}$				
	$\{A.5,B.1\}$	$\{A.4,C.2\}$				
Participation index	3/4	4/5	1/5	1/4	2/4	1/3

negative co-location can be calculated. The detailed candidate negative co-locations and participation index are listed in Table 6.2. Set the minimum frequency threshold of negative co-locations to 0.6. The participation indexes of $\{A,\bar{D}\}$, $\{B,\bar{C}\}$, and $\{C,\bar{D}\}$ are greater than 0.6, and the participation indexes of $\{A,D\}$, $\{B,C\}$, and $\{C,D\}$ are less than 0.6, so $\{A,\bar{D}\}$, $\{B,\bar{C}\}$, and $\{C,\bar{D}\}$ are prevalent co-locations.

In Figure 6.1, $\{A,B\}$, $\{A,C\}$, $\{A,\bar{D}\}$, $\{B,\bar{C}\}$, and $\{C,\bar{D}\}$ are neighbor pairs. Next, *Lemma 5* and lemmas presented by Jiang et al. (2010) are combined to generate high-size candidate negative co-locations. This method for generating candidate

TABLE 6.2
Size 2 Candidate Negative Co-Locations and Their Participation Indexes

Non-prevalent co-locations	Candidate negative co-locations	Participation index
$\{A, D\}$	$\{A, \bar{D}\}$	4/5
	$\{\bar{A}, D\}$	1/2
$\{B, C\}$	$\{B, \bar{C}\}$	3/4
	$\{\bar{B}, C\}$	2/3
$\{B, D\}$	$\{B, \bar{D}\}$	2/4
	$\{\bar{B}, D\}$	0
$\{C, D\}$	$\{C, \bar{D}\}$	2/3
	$\{\bar{C}, D\}$	1/2

co-locations is similar to the join strategy in a join-based algorithm. There are two kinds of combination methods. One is the combination of prevalent positive co-locations and prevalent negative co-locations. For example, the combination of $\{A, B\}$ and $\{A, \bar{D}\}$ can generate the candidate negative co-location $\{A, B, \bar{D}\}$. Another way is to combine prevalent negative co-locations. For example, the combination of $\{A, \bar{D}\}$ and $\{C, \bar{D}\}$ can generate the candidate negative co-location $\{A, C, \bar{D}\}$. It should be noted that two size k prevalent patterns can be combined only if they contain $k - 1$ identical instances, that is, only one instance of the two patterns is different. According to *Lemma 5* proposed in this chapter, after the candidate negative co-locations are generated, we can judge whether they are prevalent or not without another calculation of their participation indexes. For example, because $\{A, \bar{D}\}$ is a prevalent negative co-location pattern, $\{A\}$ is a subset of $\{A, C\}$ and $\{A, C\}$ is a prevalent co-location pattern; according to *Lemma 5*, the candidate negative co-location pattern $\{A, C, \bar{D}\}$ is obviously prevalent (Figure 6.2).

6.3.2 Pruning

Lemma 4 and *Lemma 5* provide effective pruning means for generating prevalent negative co-location patterns. By using *Lemma 4* and *Lemma 5*, we can quickly judge whether a candidate negative co-location pattern is prevalent, and we do not need to calculate their participation indexes one by one. For example, if $\{A, \bar{C}\}$ and $\{B, \bar{C}\}$ are the size 2 prevalent co-location patterns, and $\{A, B\}$ is the prevalent co-location pattern, then the negative co-location pattern $\{A, B, \bar{C}\}$ must be a prevalent negative co-location pattern. For example, according to *Lemma 5*, it can be seen that the participation index of candidate negative co-location pattern $\{A, B, \bar{C}\}$ is not less than participation indexes of $\{A, \bar{C}\}$ and $\{B, \bar{C}\}$ (that is, $PI(\{A, B, \bar{C}\}) \geq PI(\{A, \bar{C}\})$,

$PI(\{A,B,\bar{C}\}) \geq PI(\{B,\bar{C}\})$), since $\{A,B,\bar{C}\}$ is generated by combination of $\{A,\bar{C}\}$ and $\{B,\bar{C}\}$. There is no need to calculate the participation index of $\{A,B,\bar{C}\}$. If the co-location pattern $\{A,B\}$ is prevalent, then we can infer that the negative co-location pattern $\{A,B,\bar{C}\}$ must be a prevalent negative co-location pattern. Therefore, only the participation index of co-location $\{A,B\}$ needs to be calculated.

6.4 JOIN-BASED PREVALENT NEGATIVE CO-LOCATION PATTERNS

In this section, the join-based prevalent negative co-location pattern algorithm is presented, which includes the following steps:

1. Calculate the positive co-location pattern of all instances and use any algorithm for mining prevalent positive co-location. Store all of the size 2 prevalent co-location patterns and the PI values for all of the size 2 co-location patterns.
2. Compare the PI values of the size 2 co-location pattern with the threshold value of *min_prev* and find and store all of the size 2 candidate negative co-location patterns. The size 2 prevalent negative co-location pattern is calculated to facilitate pruning.
3. A size 2 prevalent co-location or candidate negative co-location is joined to a size 2 prevalent co-location to generate a size 3 candidate negative co-location. Then a size 3 prevalent co-location or candidate negative co-location is joined to a size 2 prevalent co-location to generate a size 4 candidate negative co-location, and so on.
4. The candidate negative co-location patterns that have been obtained from Step 1 through Step 3 are pruned. According to *Lemma 2* and *Lemma 3*, if the size 2 candidate negative co-location pattern joined is a prevalent negative co-location pattern, then it is proven to be a prevalent negative co-location pattern. The rest of the unpruned candidate negative co-location patterns are judged by the comparison between the PI value and the given threshold of *min_prev* to obtain the prevalent negative co-location patterns.

The algorithm above is summarized here.

ALGORITHM 6.1
Join-Based Prevalent Negative Co-location Pattern

Input:

1. Collection of spatial features $F = \{f_1, f_2, f_3, \cdots, f_n\}$
2. Set of spatial instances $S = \{S_1, S_2, S_3, \cdots, S_m\}$
3. Co-location relationship *R*
4. Minimum PI threshold *min_prev*

Output:

5. Size n prevalent positive co-location collection nPPC
6. Size 2 candidate negative co-location collection 2CNC
7. Size 2 prevalent negative co-location collection 2PNC
8. Size n candidate negative co-location collection nCNC
9. Size n prevalent negative co-location collection nPNC

Variable:

10. Instance co-location relation NT

Steps:

11. Calculate all NT.
12. Any existing algorithm is used to mine the set of prevalent positive co-location patterns of each size.

$$\text{Nppc} = \{nPPC_1, nPPC_2, nPPC_3, \cdots, nPPC_m\}$$

13. FOR EACH $2CNC$ & $2PPC$
14. According to *Lemma 4*, it is combined with 2PPC to generate a size 3 candidate negative co-location pattern. Eliminate repetitive patterns. The candidate negative co-location of the size 2 requires the inclusion of both negative modes.
15. FOR EACH 3 CNC & $3PPC$
16. According to *Lemma 4*, it is combined with 2PPC to generate a size 4 candidate negative co-location pattern. Eliminate repetitive patterns.
17. In addition, and so on $\cdots$
18. FOR EACH $(n-1)\,CNC$ & $(n-1)\,PPC$.
19. According to *Lemma 4*, it is combined with 2PPC to generate size n candidate negative co-location T.
20. Pruning. If T contains a prevalent low-order negative co-location, then T is a prevalent negative co-location.
21. The remaining ones without pruning are compared with the threshold *min_prev*.
22. Return to the size N prevalent negative co-location nPNC.

6.5 EXPERIMENT AND ANALYSIS

To evaluate the effectiveness of the join-based prevalent negative co-location pattern algorithm proposed in this chapter, *Algorithm 6.1* proposed in this chapter is compared with the algorithm presented in Jiang et al. (2010), which is referenced as the traditional algorithm. The algorithms are written in Python, and the experimental environment is PyCharm running in Windows10.

6.5.1 DATA SETS

The real data set selected in the experiment is the geospatial data from Shopping, Traffic, Dining and Companies (see Table 6.3) in the City of Ji'nan, Shandong, China, which contains a total of 11,189 data points, as shown in Figure 6.3.

6.5.2 JOIN-BASED PREVALENT NEGATIVE CO-LOCATION PATTERNS

According to the obtained data, a positive co-location of size 1–4 is obtained, as shown in Figure 6.4. The line charts represent the sum of the total number of positive

TABLE 6.3
Data Set

Type	Abbreviation	Number
Shopping	S	7,284
Traffic	T	582
Dining	D	1,963
Companies	C	1,360

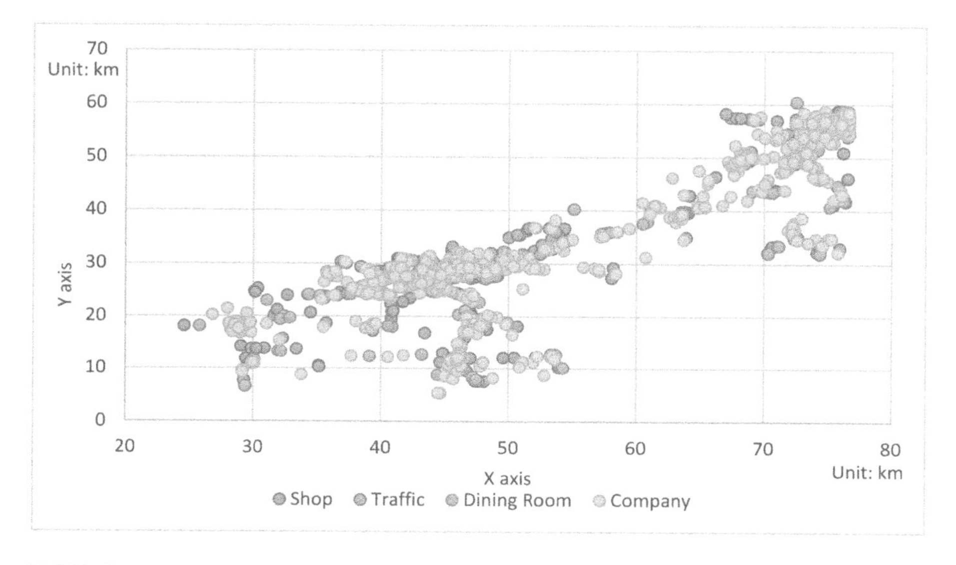

FIGURE 6.3 The data set of City of Ji'nan, Shandong, China.

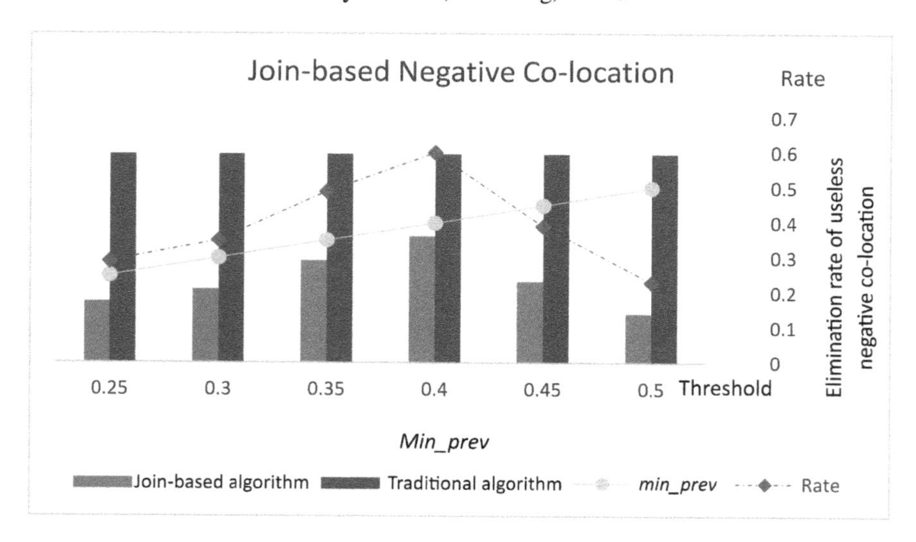

FIGURE 6.4 A comparison between the negative co-location algorithm based on the join-based and traditional algorithms with the change in threshold value.

co-location patterns for each size, and the bar charts represent the number of detailed positive co-location patterns for each size.

In the experiment, the distance threshold is fixed at d = 1,000 m. The participation threshold was changed from 0.1 to 0.7, and the relevant tests were performed. In addition, the traditional mining algorithm and the algorithm proposed in this chapter are used to mine all negative co-locations under different thresholds.

As seen in Figure 6.4, the effect of the connection algorithm changes with the *min_prev*. When the number of the size 2 prevalent co-locations is equal to the number of size 2 candidate negative co-locations, it is the most complicated case, and the effect is the worst. However, its number is approximately 0.6 of the traditional algorithm. Moreover, the traditional algorithm enumerates all negative co-location patterns, independent of the value of *min_prev*, without fluctuations.

6.5.3 Difficulties in Mining Negative Co-Location Patterns

Jiang et al. (2010) has summarized the difficulties in mining negative co-location patterns. Combining these with our experiments, the major difficulties for mining negative co-location patterns are:

1. The spatial computation is time consuming, including generation of the negative co-location patterns. The number of candidate negative co-location patterns exponentially increases with the size of data sets (Jiang et al. 2010).
2. The algorithms developed for mining the negative association rules cannot be reused; also, the algorithms developed for mining positive co-location mining cannot be used directly since some methods must be redesigned; there are no traditional "transactions" in spatial data sets (Jiang et al. 2010).

6.6 CONCLUSIONS

In summary, the main contributions of this chapter can be summarized as follows:

- A candidate negative co-location pattern is proposed based on the definition of prevalent negative co-location, and it is proven that any N size candidate negative co-location pattern can be joined by an $N - 1$ size prevalent co-location pattern or prevalent negative co-location pattern and a size 2 prevalent co-location pattern.
- For the specified spatial feature set $\overline{Y}$, the negative co-location pattern $T = X \cup \overline{Y}$ of the specified size can be calculated directly through the join-based algorithm.
- According to the definition of a negative co-location pattern, the monotonous non-decrement of the PI value of a negative co-location pattern is strictly proven, and a fast pruning method is proposed by using the corresponding lemma.
- By combining negative co-location patterns from low order to high order, two patterns in extreme cases and their meanings are proposed: "single positive of negative co-location" and "single negative of negative co-location." An algorithm for solving the pattern is given.

REFERENCES

Bao, X., and Wang, L., A clique-based approach for co-location pattern mining. *Information Sciences*, vol. 490, 2019, pp. 244–264.

Cao, L., Dong, X., and Zheng, Z. e-NSP: Efficient negative sequential pattern mining. *Artificial Intelligence*, vol. 235, 2016, pp. 156–182.

Dong, X., Gong, Y., and Cao, L., F-NSP+: A fast negative sequential patterns mining method with self-adaptive data storage. *Pattern Recognition*, vol. 84, 2018, pp. 13–27.

He, F.Z., Jia, Z.Y., and Zhang, D.D., Mining spatial co-location pattern based on parallel computing. *Journal of Yunnan Normal University (Natural Science Edition)*, vol. 35, no. 4, 2015, pp. 56–62.

Huang, Y., Shekhar, S., and Xiong, H., Discovering colocation patterns from spatial data sets: A general approach. *IEEE Transactions on Knowledge and Data Engineering*, vol. 16, no. 12, 2004, pp. 1472–1485.

Jiang, Y., Wang, L., and Chen, H., Discovering both positive and negative co-location rules from spatial data sets. Software Engineering and Data Mining (SEDM). *Proceedings of the 2010 2nd International Conference on IEEE*, Chengdu, China, June 23–25, 2010, pp. 398–403. *IEEE International Conference on Data Mining*, Omaha, NE, USA, October 28–31, 2007.

Jiang, Y., Wang, L., Lu, Y., et al., Discovering both positive and negative co-location rules from spatial data sets. *Proceedings of the 2nd International Conference on Software Engineering and Data Mining*, Chengdu and Piscataway: IEEE Press, June 23–25, 2010, pp. 398–403.

Morimoto, Y., Mining Frequent Neighboring Class Sets in Spatial Databases. *Proc. of the seventh ACM SIGKDD Int. Conf. on Knowledge Discovery and Data Mining*, San Francisco, California, 2001, pp. 353–358.

Shekhar, S., and Huang, Y., Co-location rules mining: A summary of results. *Proc. of International Symposium on Spatio and Temporal Database (SSTD'01)*, Springer-Verlag, Berlin, Heidelberg, New York, 2001, pp. 236–240.

Wang, L., Bao, X., Chen, H., et al., Effective lossless condensed representation and discovery of spatial co-location patterns. *Information Sciences*, vol. 436–437, 2018, pp. 197–213.

Wang, L., Bao, X., and Zhou, L., Redundancy reduction for prevalent co-location patterns. *IEEE Transactions on Knowledge and Data Engineering*, vol. 30, no. 1, 2018, pp. 142–155.

Wang, L., Bao, Y., and Lu, Z., Efficient discovery of spatial co-location patterns using the iCPI-tree. *The Open Information Systems Journal*, vol. 3, no. 2, 2009, pp. 69–80.

Wang, L., Bao, Y., Lu, J., and Yip, J., A new join-less approach for co-location pattern mining. *Proc. of the IEEE 8th International Conference on Computer and Information Technology (CIT2008)*, Sydney, Australia, 2008, pp. 197–202.

Wang, L., Zhou, L., Lu, J., et al., An order-clique-based approach for mining maximal co-locations. *Information Sciences*, vol. 179, no. 19, 2009, pp. 3370–3382.

Wu, X., Zhang, C., and Zhang, S., Efficient mining of both positive and negative association rules. *ACM Transactions on Information Systems*, vol. 22, no. 3, 2004, pp. 381–405.

Yoo, J.S., Shekhar, S., and Celik, M., A join-less approach for co-location pattern mining: A summary of results. *Proceedings of the IEEE International Conference on Data Mining*, Houston, Piscataway: IEEE Press, November 27–30, 2005, pp. 813–816.

Yoo, J.S., Shekhar, S., Smith, J., et al., A partial join approach for mining co-location patterns. *Proceedings of the 12th Annual ACM International Workshop on Geographic Information Systems (GIS)*, Washington and New York: ACM Press, November 12–13, 2004, pp. 241–249.

Zheng, Z., Zhao, Y., Zuo, Z., et al., Negative-GSP: An efficient method for mining negative sequential patterns. *Proceedings of the Eighth Australasian Data Mining Conference*, Melbourne, December, New York: ACM Press, 2009, pp. 63–67.

Zhou, G., Co-location decision tree for enhancing decision-making of pavement maintenance and rehabilitation. *Ph.D. dissertation*, Virginia Tech, Blacksburg, VI, USA, 2011.

Zhou, G., Huang, S., Wang, H., Zhang, R., Wang, Q., Sha, H., Liu, X., and Pan. Q., A buffer analysis based on co-location algorithm. *ISPRS: International Archives of the Photogrammetry, Remote Sensing and Spatial Information Sciences*, XLII-3, 2018, pp. 2487–2490.

Zhou, G., Li, Q., and Deng, G.M., Maximal instance algorithm for fast mining of spatial co-location patterns: Patterns. *Remote Sensing*, vol. 13, 2021, p. 960.

Zhou, G., Li, Q., Deng, G.M., Yue, T., and Zhou. X., Mining co-location patterns with clustering items from spatial data sets. *ISPRS: International Archives of the Photogrammetry, Remote Sensing and Spatial Information Sciences*, XLII-3, 2018, pp. 2505–2509.

Zhou, G., Shi, Y., and Zhang. R., Co-location decision tree model for extracting exposed carbonate rocks in karst rocky desertification area, SPIE. *Asia-Pacific Remote Sensing*, Beijing, October 13–16, 2014.

Zhou, G., and Wang, L.B., Co-location decision tree for enhancing decision-making of pavement maintenance and rehabilitation. *Transportation Research Part C*, vol. 21, 2012, pp. 287–305.

Zhou, G., Wei, J., Zhou, X., Zhang, R., Huang, W., Sha, H., and Chen, J., An isometric mapping based co-location decision tree algorithm. *ISPRS: International Archives of the Photogrammetry, Remote Sensing and Spatial Information Sciences*, XLII-3, 2018, pp. 2531–2534.

Zhou, G., Zhang, R.T., Zhang, D.J., Norman, K., Soe, M., Clement, A., and Prasad, S., Thenkabail: Manifold learning co-location decision tree for remotely sensed imagery classification. *Remote Sensing*, vol. 8, no. 10, 2016, p. 855.

7 Application of Mining Co-Location Patterns in Pavement Management and Rehabilitation

7.1 INTRODUCTION

7.1.1 Distress Rating

As an initial experimental study, this chapter will first explore the decision tree and decision rules induction using the following nine common types of pavement distresses, which are listed in Table 7.1 (Zhou and Wang 2010, 2012; Zhou 2011):

- Alligator Cracking,
- Block Cracking,
- Transverse Cracking,
- Bleeding,
- Rutting,
- Utility Cut Patching,
- Patching Deterioration, and
- Raveling.

Table 7.1 presents the eight types of distresses that are evaluated for asphaltic concrete pavements. In Table 7.1, the severity of distress is rated in four categories, ranging from very slight to very severe. Extent (or density) is classified in five categories, ranging from few (less than 10%) to throughout (more than 80%). The identification and description of distress types, severity, and density are:

- The road conditions of Alligator Cracking are rated as a percentage of the section that falls under the categories of None, Light, Moderate, and Severe. Percentages are shown as 1 = 10%, 2 = 20%, and 3 = 30%, up to 10 = 100%. The appropriate percentages should be placed under None, Light, Moderate, and Severe. These percentages should always add up to 100%.
- The severity levels of distresses Block Cracking, Transverse Cracking, Bleeding, Rutting, Utility Cut Patching, Patching Deterioration, and Raveling are rated at 4 levels: None (N), Light (L), Moderate (M), and Severe (S), respectively.
- The severity levels of ride quality are classified as Average (L), Slightly Rough (M), and Rough (S).

DOI: 10.1201/9781003139416-7

TABLE 7.1
Nine Common Types of Distresses for This Study

Distress	Rating	
Alligator Cracking	Alligator None (AN) Alligator Light (AL) Alligator Moderate (AM) Alligator Severe (AS)	Percentages of 1 = 10%, 2 = 20%, 3 = 30%, up to 10 = 100% indicate None, Light, Moderate, and Severe, respectively
Block/Transverse Cracking (BK)	This indicates the overall condition of the section as follows: • N-None • L-Light • M-Moderate • S-Severe	
Reflective Cracking (RF)	The same manner as BK	
Rutting (RT)	The same manner as BK	
Raveling (RV)	The same manner as BK	
Bleeding (BL)	The same manner as BK	
Patching (PA)	The same manner as BK	
Utility Cut Patching,	The same manner as BK	
Ride Quality (RQ)	The condition is designated as follows: • L–Average • M–Slightly Rough • S–Rough	

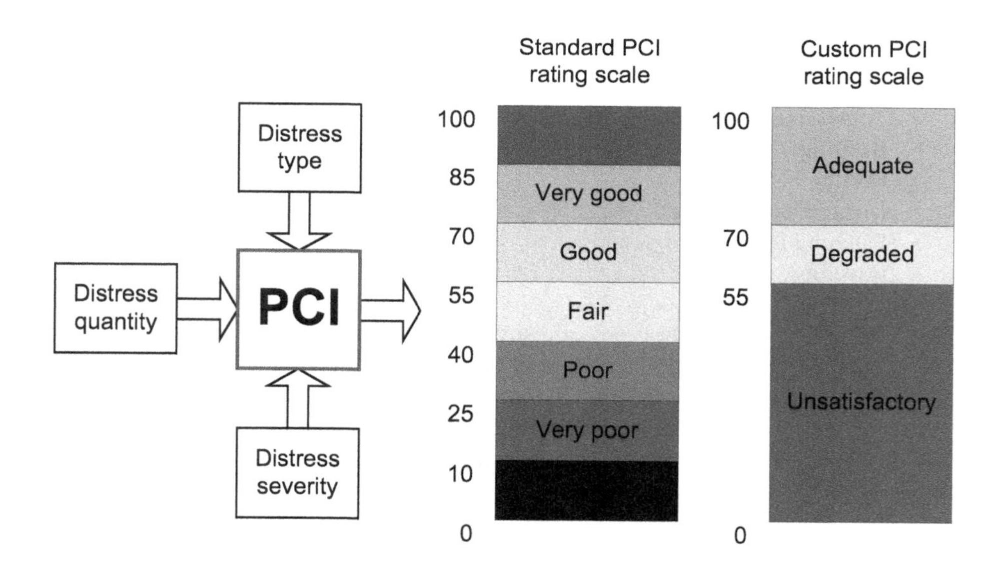

FIGURE 7.1 Pavement Condition Index standard and custom rating scales.
Source: Courtesy of Greene, J. and M. Shahin, 2010.

Experiments, accompanying all the analyses and pavement condition evaluations presented in this chapter, are based on the pavement performance measures. A common acceptable pavement performance measure is the Pavement Condition Index (PCI), which was first defined by the US Army (see Figure 7.1). In the PCI, the pavement condition is related to factors such as structural integrity, structural capacity, roughness, skid resistance, and rate of distress. These factors are quantified in the evaluation worksheet that field inspectors use to assess and express the local pavement condition and damage severity. Mostly, inspectors use their own judgment to assess the distress condition. Usually, the PCI is quantified into seven levels, from Excellent (over 85) to Failed 0 (see Figure 7.1). Thus, PCI is an important index for maintenance and repair determination in which the overall conditions of the observed road surface are evaluated.

7.1.2 Potential Rehabilitation Strategies

Based on the knowledge gained from experts, we have classified rehabilitation treatments for flexible pavements into three main categories according to the type of the problem to be corrected: cracking, surface defect problems, and structural problems. These problems can be treated using crack treatment, surface treatment, and non-structural overlay (one- and two-course overlay), respectively.

In order to select an appropriate treatment for rehabilitation and maintenance to a specific road, seven potential rehabilitation and maintenance strategies have been proposed by the North Carolina Department of Transportation (NCDOT) (Table 7.2). Which treatment strategies will be carried out for a pavement segment is dependent on the comprehensive evaluation of all distresses. This used to be created by experts or a pavement engineer at North Carolina Department of Transportation. This research will experiment and test whether the decision tree and decision rule can produce an appropriate decision for an M&R (maintenance and repair) strategy using data mining technology and then compare the differences of decisions made by the manual method and data mining.

7.2 EXPERIMENTAL DESIGN

7.2.1 Flowchart of Experiment and Comparison Analysis

The proposed co-location decision tree algorithm consists of two major steps, co-location mining and decision tree induction, as depicted in Figure 7.2. When a database is input, the spatial data and nonspatial data are selected. Both types of data are

TABLE 7.2
Potential Rehabilitation Strategies

ID	Rehabilitation Strategies
0	Nothing
1	Crack Pouring (CP)
2	Full-Depth Patch (FDP)
3	1" Plant Mix Resurfacing (PM1)
5	2" Plant Mix Resurfacing (PM2)
6	Skin Patch (SKP)
7	Short Overlay (SO)

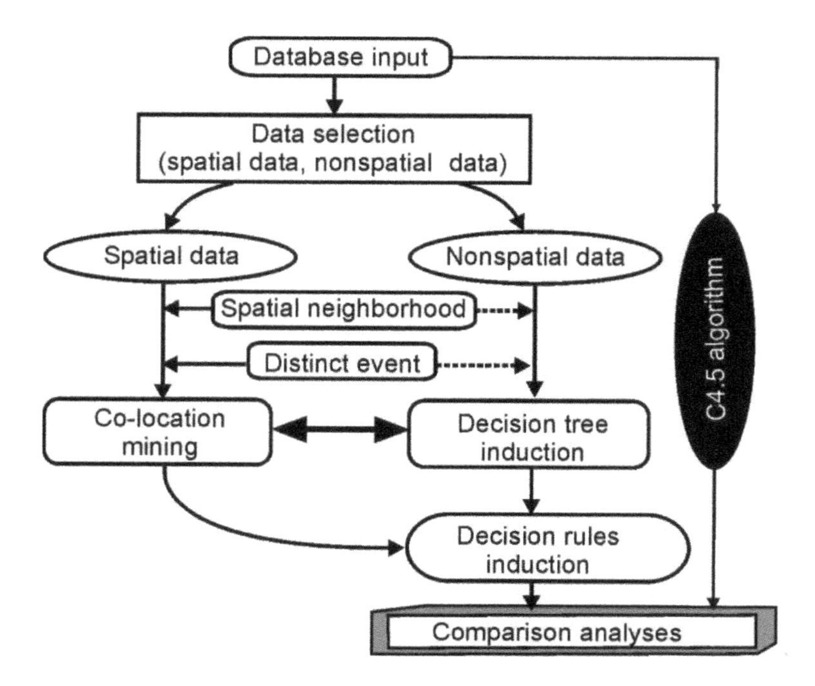

FIGURE 7.2 Flowchart of experimental design.

employed to induce co-location rules, and only nonspatial data is used for decision tree induction. The decision tree induction algorithm is C4.5. The induced co-location rules are used for determining pruning (merging) the branches of a tree. Thus, the process is recursively repeated until the leaf node creation of a tree. Following up on the co-location decision tree induction, the co-location decision rules are induced by translating a decision tree into semantic expressions. The comparison analyses for three methods are conducted to evaluate the advantages and disadvantages of the proposed CL-DT algorithm.

7.2.2 Data Sources

In 1983, the Institute for Transportation Research and Education (ITRE) of North Carolina State University began working with the Division of Highways of the North Carolina Department of Transportation (NCDOT) to develop and implement a Pavement Management System for its 60,000 miles of paved state highways. At the request of several municipalities, NCDOT has made this Pavement Management System available for North Carolina municipalities. The ITRE modified this system for municipal streets in more than 100 municipalities in North and South Carolina. The data sources for this experiment are provided by ITRE of North Carolina State University. They conducted pavement distress surveys for several counties since January 2007 to determine whether or not the activity (rehabilitation treatment) for pavement needs to be carried out. The collected 1,285 records to be utilized in this empirical study come from a network-level survey covering several-county roads including US Highway 1 and the rural road network. The provided pavement database is a spatial-based rational database, that is, an ArcGIS software compatible database. In this database, 89 attributes including geospatial attributes (e.g., X,Y coordinates, central line, width of lane, etc.), pavement condition attributes (e.g., cracking, rutting, etc.), traffic attributes (e.g., shoulder, lane number, etc.), and economic attributes (e.g., initial cost, total cost) are recorded by engineers, who were carrying out these surveys by walking or driving and recording the distress information and their corresponding M&R strategy. Then the data is integrated into database, as shown in Figure 7.3. The first through 19th column recorded road name, type, class, owner, etc. attributes; the 31st through 43rd column recorded the pavement condition (distress) attributes; the 44th through 50th column recorded the different types of cost; and others included the proposed activities, etc.

7.2.3 Nonspatial Attribute Data Selection

Which nonspatial data are more significant to contribute the decision making of pavement maintenance and rehabilitation? From the expert's suggestion in the ITRE, the following nine common types of distresses are considered to assess the necessity for road rehabilitation (see Table 7.3). The rating was determined by visual evaluation/inspection at each section of roadway. The severity of distresses in Table 7.3 is rated in four categories ranging from very slight to very severe. Extent (or density) is also classified in five categories ranging from few (less than 10%) to throughout (more than 80%). The identification and description of types of distress, severity, and density are as follows, respectively.

- The road conditions of the Alligator Cracking are rated as a percentage of the section that falls under the categories of None, Light, Moderate, and Severe. Percentages are shown as 1 = 10%, 2 = 30%, 3 = 60%, up to 10 = 100%. The appropriate percentages were placed under None, Light, Moderate, and Severe. These percentages should always add up to 100%.
- The severity levels of distresses for Block Cracking, Transverse Cracking, Bleeding, Rutting, Utility Cut Patching, Patching Deterioration, and Raveling are rated four levels: None (N), Light (L), Moderate (M), and Severe (S), respectively.
- The severity levels of ride quality are classified as: Average (L), Slightly Rough (M), and Rough (S).

7.2.4 Spatial Attribute Data Selection

Traditionally, only nonspatial attribute data were considered for mining decision tree. However, as mentioned early, it is incorrect for decision making without considering spatial data. The spatial data in the database include X,Y coordinates, central line, width of lane, number of travel lanes, length of street segment, first-left, to-left,

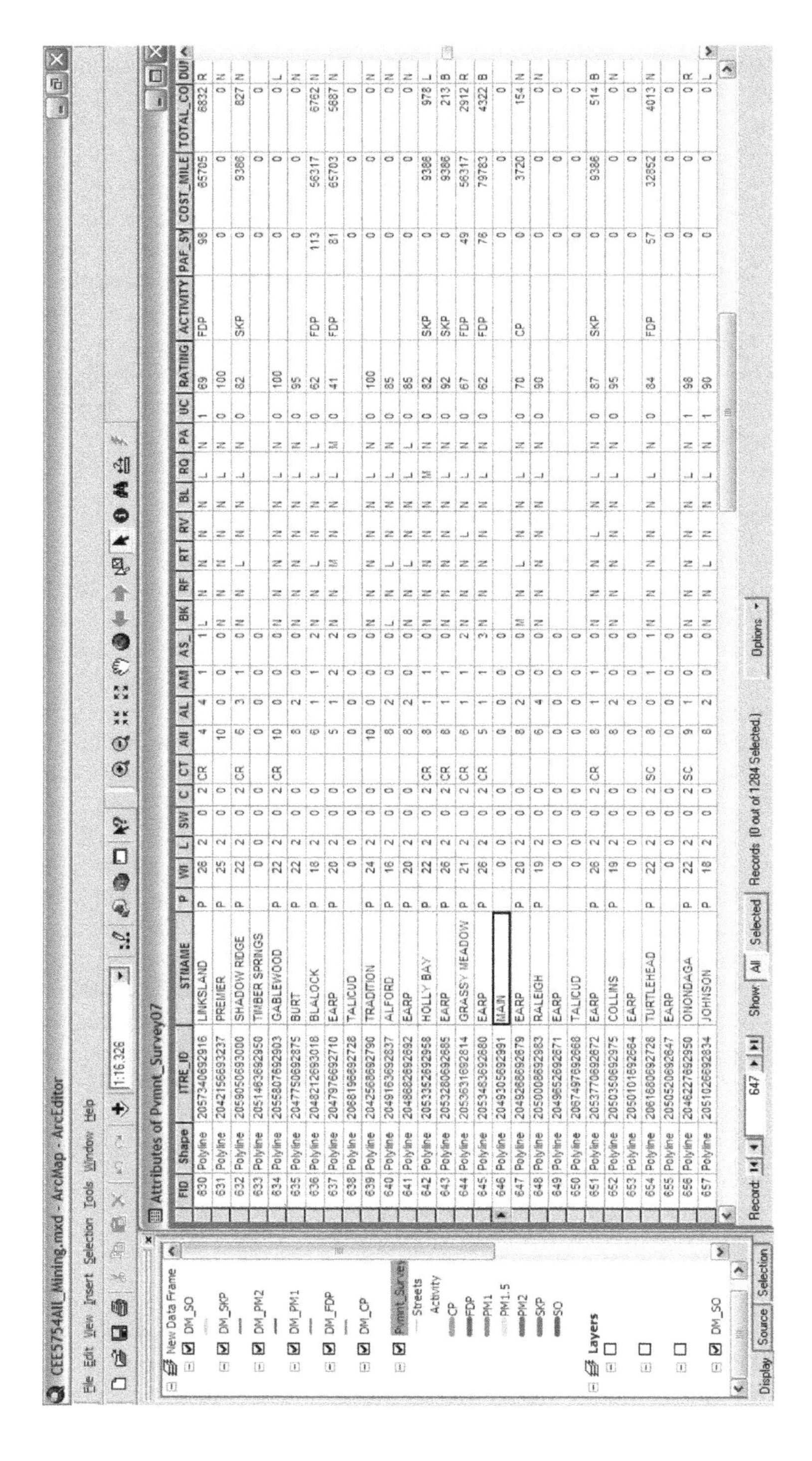

FIGURE 7.3 The spatial-based pavement database.

TABLE 7.3
Eight Common Types of Distresses Plus Ride Quality for This Study

#	Distress	Rating	
1	Alligator Cracking (four types of rates are given)	Alligator None (AN) Alligator Light (AL) Alligator Moderate (AM) Alligator Severe (AS)	Percentages of 1 = 10%, 2 = 20%, 3 = 30%, up to 10 = 100% indicate None, Light, Moderate, and Severe, respectively
2	Block/Transverse Cracking (BK)	This indicates the overall condition of the section as follows: • N-None • L-Light • M-Moderate • S-Severe	
3	Reflective Cracking (RF)	The same rating as BK's	
4	Rutting (RT)	The same rating as BK's	
5	Raveling (RV)	The same rating as BK's	
6	Bleeding (BL)	The same rating as BK's	
7	Patching (PA)	The same rating as BK's	
8	Utility Cut Patching	The same rating as BK's	
9	Ride Quality (RQ)	The condition is designated as follows: • L–Average • M–Slightly Rough • S–Rough	

TABLE 7.4
Selected Spatial Data for This Study

Attributes	Explanation
• X coordinate • Y coordinate	Datum: NAD_1983_StatePlane_North_Carolina_FIPS_3200_Feet Coordinate system name: GCS_North_American_1983 Map Projection Name: Lambert Conformal Conic Standard Parallel: 34.333333 Standard Parallel: 36.166667 Longitude of Central Meridian: –79.000000 Latitude of Projection Origin: 33.750000 False Easting: 2000000.002617 False Northing: 0.000000
Length	GIS length of street segment (in feet)

first-right, to-right, etc. However, only two types of *spatial data*, XY coordinates, and length of each road segments are considered (see Table 7.4).

7.2.5 Maintenance and Rehabilitation (M&R) Strategies

Based on the knowledge obtained from ITRE in the field, the ITRE has classified flexible pavement rehabilitation needs into three main categories according to the type of the problem to be corrected: (1) cracking, (2) surface defect problems, and (3) structural problems. These problems can be treated using crack treatment, surface

TABLE 7.5
Potential Rehabilitation Strategies

#	Rehabilitation Strategies
0	Nothing
1	Crack Pouring (CP)
2	Full-Depth Patch (FDP)
3	1" Plant Mix Resurfacing (PM1)
4	2" Plant Mix Resurfacing (PM2)
5	Skin Patch (SKP)
6	Short Overlay (SO)

treatment, and nonstructural overlay (one-and two-course overlay), respectively. In order to select an appropriate pavement treatment, seven potential rehabilitation and maintenance strategies have been proposed by the NCDOT (see Table 7.5). Which treatment strategies will be carried out for a pavement segment is traditionally determined by the ITRE, who comprehensively evaluate all types of distresses. This chapter intends to replace this process using the CL-DT algorithm.

7.3 INDUCTION OF CO-LOCATION MINING RULES

7.3.1 Determination of Candidate Co-Locations

As described in Chapter 3, the candidate instances with co-location relationship will be determined using the spatial neighborhood criterion with a given threshold, D_θ. In this research, the spatial neighborhoods for all instances are computed by:

$$Dist_{i,j} = \sqrt{(X_i - X_j)^2 + (Y_i - Y_j)^2} \qquad \forall i,j = 1,2,\cdots,1285 \tag{7.1}$$

Where X and Y are the spatial data of the pavement database. With the given database at a dimension of 1,285 instances, the spatial distances of any two instances produce a matrix with the dimension of 1,285 × 1,285, i.e.,

$$\underset{1285\times1285}{Dist} = \begin{bmatrix} 0 & d_{12} & \cdots\cdots & d_{1\times1285} \\ & 0 & \cdots\cdots & d_{2\times1285} \\ & & \cdots\cdots & \cdots \\ & & \cdots & d_{1284\times1285} \\ & & & 0 \end{bmatrix} \tag{7.2}$$

With the given database, a statistical analysis, including average and standard deviation, for the length of street segment is conducted. It is found that the length of approximately 25,000 feet is appropriate as threshold. Thus, the threshold of spatial distance of two instances is selected $D_\theta = 25,000$ (feet). Combining the generated spatial neighborhood matrix (equation 7.2) and threshold, the elements of spatial neighborhood matrix, $\underset{1285\times1285}{Dist}$, are re-calculated *by:*

$$d_{i,j} = \begin{cases} d_{i,j} & \text{if } d_{i,j} \leq 25,000 \\ 0.0 & \text{if } d_{i,j} > 25,000 \end{cases} \qquad d_{i,j} \subseteq Dist \tag{7.3}$$

With the above filtering, the potential of co-location instances can be determined by the spatial neighborhood matrix, which is a spare matrix.

7.3.2 Determination of Table Instances of Candidate Co-Locations

7.3.2.1 Determination of Distinct Events

In addition to the above geospatial distance constraint, another constraint condition for the determination of candidate co-location is the distinct event-type constraint. This implies that if two instances are co-located, they must be distinct events. The constraint condition of distinct event is mathematically expressed by:

$$\Gamma_i = \sum_{k=1}^{K} \left(\left\| f_i - v_k \right\| \right)^2 \qquad i = 1, \cdots, 1285 \tag{7.4}$$

where $\|x_i - v_k\|$ represents the Euclidean distance between f_i and v_k; $V = \{v_1, v_2, \cdots, v_k\}$ is a set of corresponding clusters center of attributes, $\{a_1, a_2, \cdots, a_k\}$; Γ is a squared error clustering criterion; and K is number of event. $v_k, \forall k = 1, 2, \cdots K$ can be calculated by:

$$v_k = \sum_{i=1}^{N} f_i / K, \qquad \forall k = 1, 2, \cdots, K \tag{7.5}$$

So the distinct event can be determined by:

$$\Gamma_i = \begin{cases} \text{event} & \text{if } \Gamma_i \leq \Gamma_\theta \\ N/A & \text{if } \Gamma_i \leq \Gamma_\theta \end{cases} \tag{7.6}$$

where Γ_θ is the threshold. If the Γ_i is greater than a given threshold, the i-th instance is assumed to be the distinct event.

For the given database, only one attribute, ride quality (RQ), is selected for evaluating the distinct event with which the clusters center of attributes of ride quality, v *is 85, which is* calculated by equation 7.5.

Equation 7.4 can be rewritten as:

$$\Gamma_i = \sum_{i=1}^{K} (f_i - 85)^2 \tag{7.7}$$

TABLE 7.6
The Process of Co-Location Mining Using Both the Geospatial Distance Criterion and Distinct Event Criterion

#	X	Y	Activity	Rating
1	2049671.1	691641.5	PM1	69
2	2049518.9	691461.1		98
3	2049600.3	691368.6		90
4	2049673.2	690247.2	CP	68
5	2049643.1	697413.1		100
6	2049600.3	691368.6		90
7	2049646.1	690634.4		88
8	2049632.7	702497.6		
9	2049615.2	699440.8		
10	2049660.7	693303.0		
11	2049652.9	692671.9		

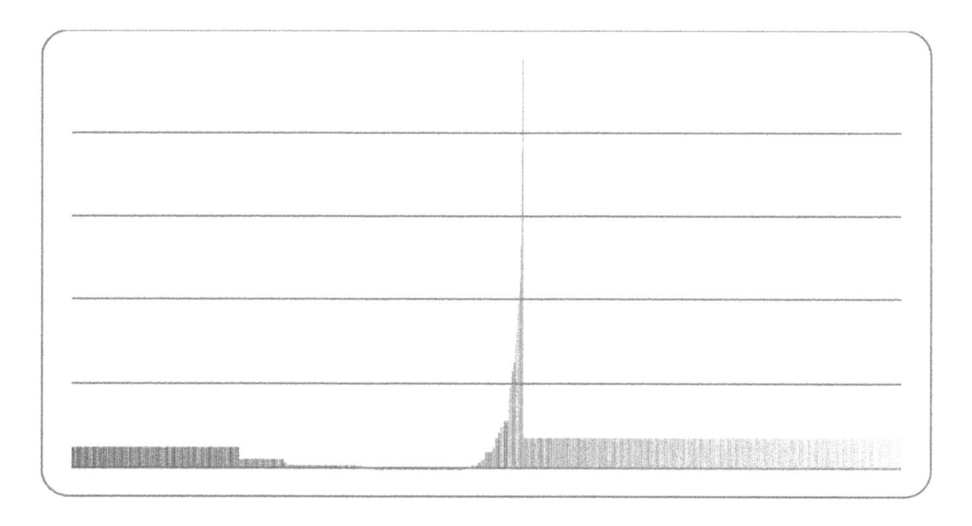

FIGURE 7.4 Determination of distinct events using ride quality (RQ).

With equation 7.7, the values of Γ_i for 1,285 instances is depicted in Figure 7.3. Further, the distinct events can be determined by equation 7.6.

The computational process of these two constrain conditions can be illustrated by Table 7.6. For example, for a given distinct event, PM1, the geospatial distance criterion first produces 11 instances, which are co-located with PM1. With the second criterion condition of distinct event, only 7 instances are co-located with PM1 event, since the other 4 instances have no records of rating of ride quality. Figure 7.4 depicts the distributions from the original 1,285 instances (Figure 7.5a) to 946 instances

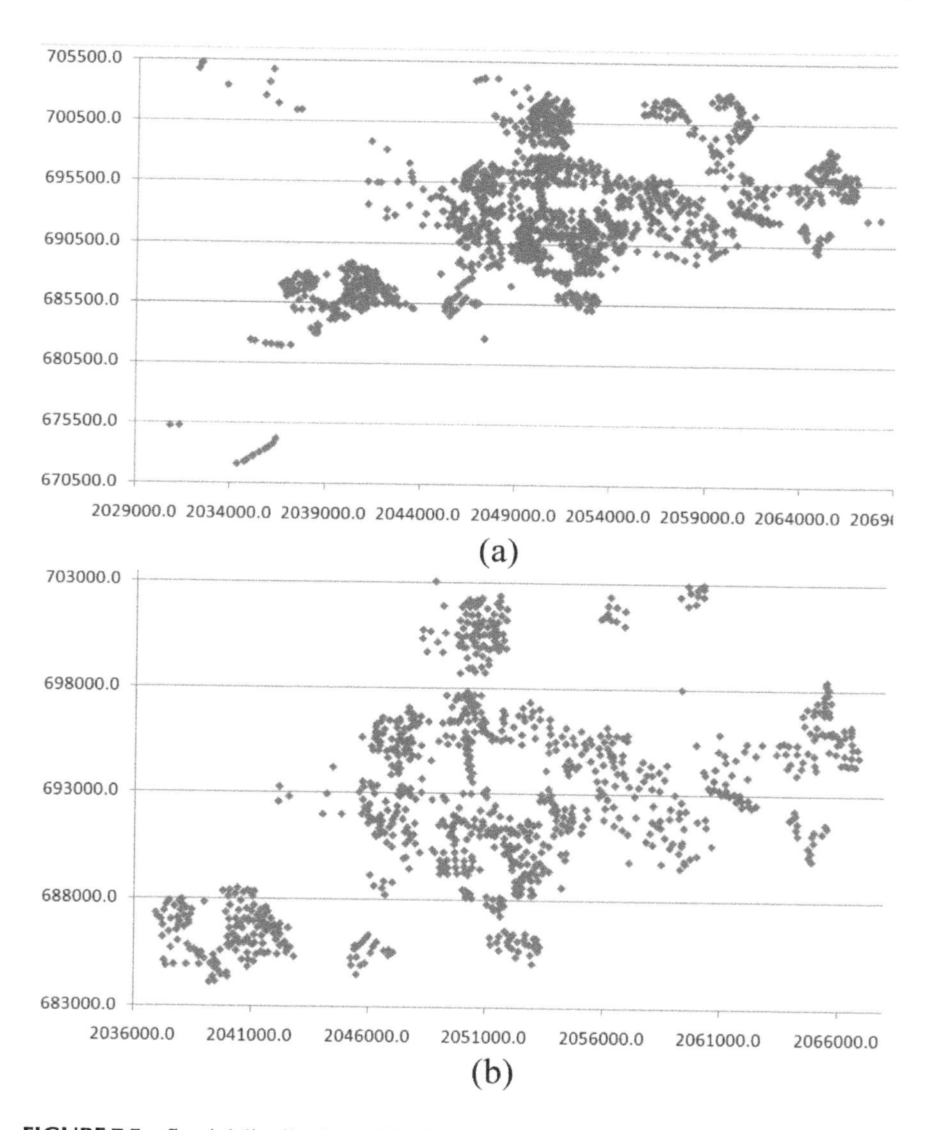

FIGURE 7.5 Spatial distributions of (a) the 1,285 original instances and (b) the 946 instances after co-location algorithm

(Figure 7.5b) after two constraint conditions are used. Finally, a total of 946 distinct events are found.

7.3.2.2 Co-Location Mining for Rehabilitation and Maintenance Strategy

As mentioned earlier, seven potential rehabilitation and maintenance strategies have been proposed by the North Carolina Department of Transportation (NCDOT). In order to find the co-location events for each M&R strategy, we take each strategy as a distinct event and then find the co-location using the co-location mining algorithm, which has been described earlier, respectively. For each of M&R treatment strategies, the results of co-locating mining are as follows.

7.3.2.2.1 Crack Pouring (CP)

The ITRC at the NCDOT indicated three crack pouring (CP) treatment strategies. Two of them are chosen to illustrate the results of the proposed co-location mining method. As seen in Figure 7.6a and Figure 7.6b, five instances are clustered with the first CP event (Figure 7.6a), and three instances are clustered with another CP event (Figure 7.6b). Other events are not clustered due to far distances.

7.3.2.2.2 Full-Depth Patch (FDP)

The ITRC at the NCDOT indicated 34 full-depth patch (FDP) treatment strategies. Four of them are chosen to illustrate the results of the proposed co-location mining algorithm. As seen in Figure 7.7a and Figure 7.7d, no instances are clustered around

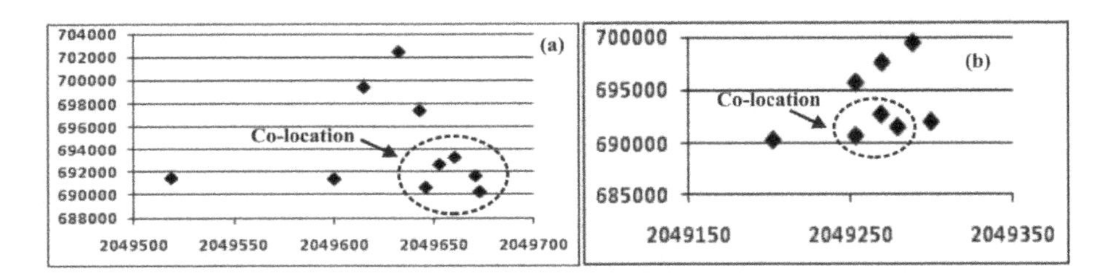

FIGURE 7.6 Spatial distributions of CP instances after initial determination of co-location algorithm.

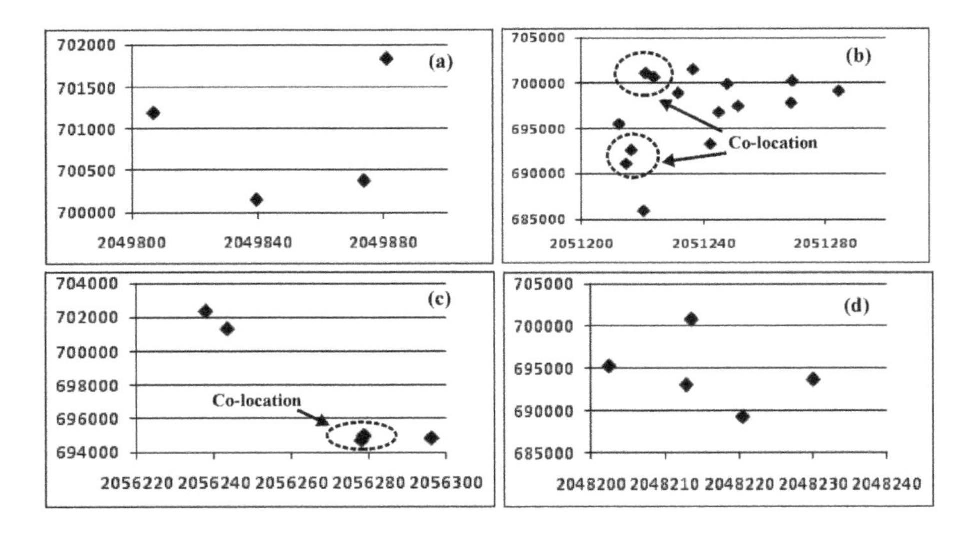

FIGURE 7.7 Spatial distributions of FDP instances after initial determination of co-location algorithm.

the two FDP events, but there are clusters in Figure 7.7b and 7.7c for the other two FDP events.

7.3.2.2.3 1" Plant Mix Resurfacing (PM1)

The ITRC at the NCDOT indicate six 1" plant mix (PM1) treatment strategies. All of them are chosen to illustrate the results of the proposed co-location mining algorithm. As seen from Figure 7.8a, Figure 7.8c, Figure 7.6d, and Figure 7.8f, no instances are clustered with the four PM1 events, but there are clusters for the two PM1 events in Figure 7.8b and 7.8e.

7.3.2.2.4 2" Plant Mix Resurfacing (PM2)

The ITRC at the NCDOT indicated three 2" plant mix (PM2) treatment strategies. All of them are chosen to illustrate the results of the proposed co-location mining algorithm on the basis of the event of PM2 treatment strategy. As seen in Figure 7.9a and Figure 7.9b, no instances are clustered with the two PM2 events, but there is a cluster for another PM2 event in Figure 7.9c.

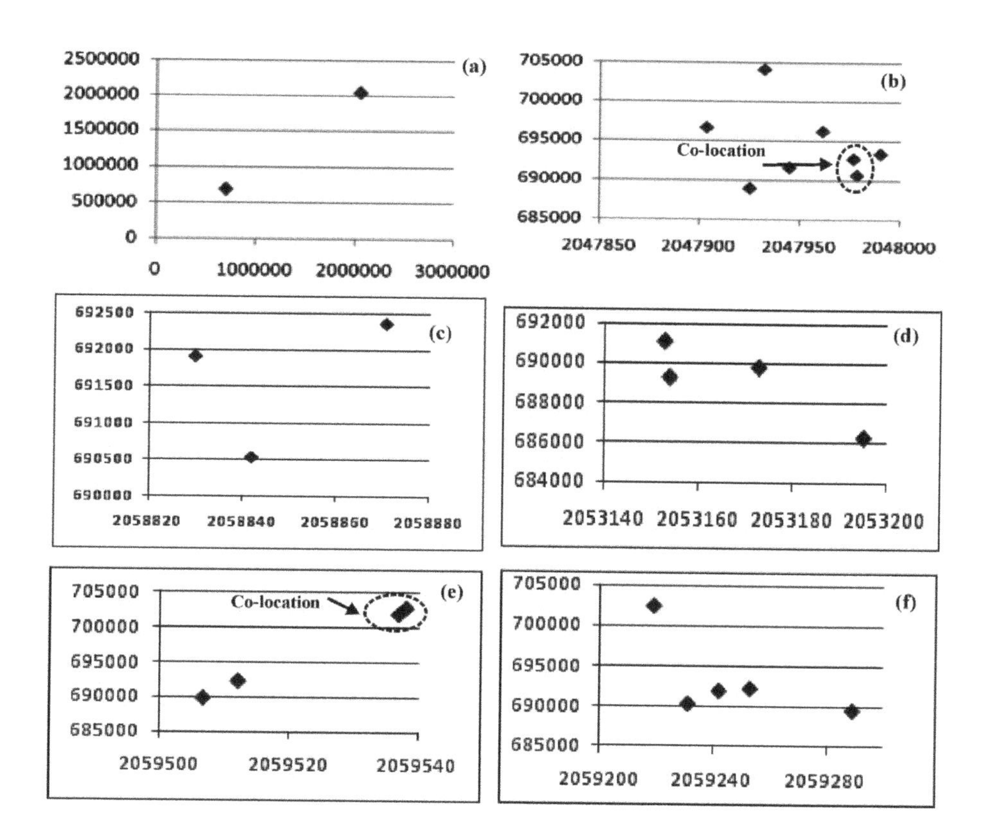

FIGURE 7.8 Spatial distributions of PM1 instances after initial determination of co-location algorithm.

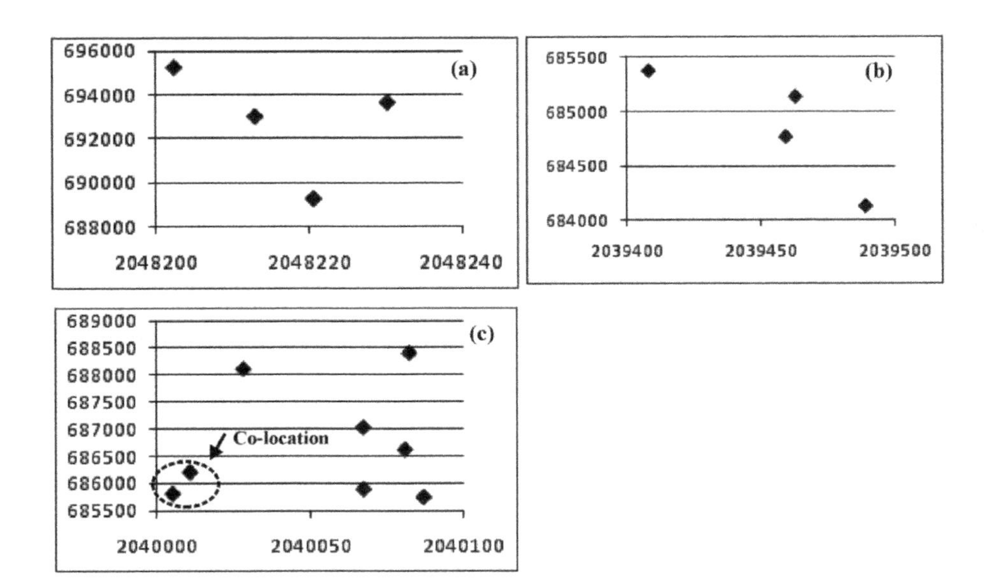

FIGURE 7.9 Spatial distributions of PM2 instances after initial determination of co-location algorithm.

7.3.2.2.5 Skin Patch (SKP)

The ITRC at the NCDOT indicated 56 skin patch (SKP) treatment strategies. Nine representatives of them are chosen to illustrate the results of the proposed co-location mining method with each SKP treatment strategy. As seen in Figure 7.10a through Figure 7.10d, Figure 7.10f, and Figure 7.10g through Figure 7.10i, several candidate events are clustered surrounding the individual SKP treatment event, but no candidate event is clustered surrounding one SKP treatment event in Figure 7.10e.

7.3.2.2.6 Short Overlay (SO)

The ITRC at the NCDOT indicated three short overlay (SO) treatment strategies. All of them are chosen to illustrate the results of the proposed co-location mining algorithm for each SO treatment strategy. As seen in Figure 7.11a and Figure 7.11b, there are clusters surrounding the SO treatment, but no cluster surrounding another SO treatment in Figure 7.11c.

7.3.2.3 Pruning

These generated candidates of co-location events for each treatment strategy may include incorrect determinations. The purpose of pruning is to remove the non-prevalent co-locations from the candidate prevalent co-location set so that the further co-location mining rule induction is reliable. To this end, a cross-correlation criterion of spatial attributes is applied to eliminate those non-prevalent co-location

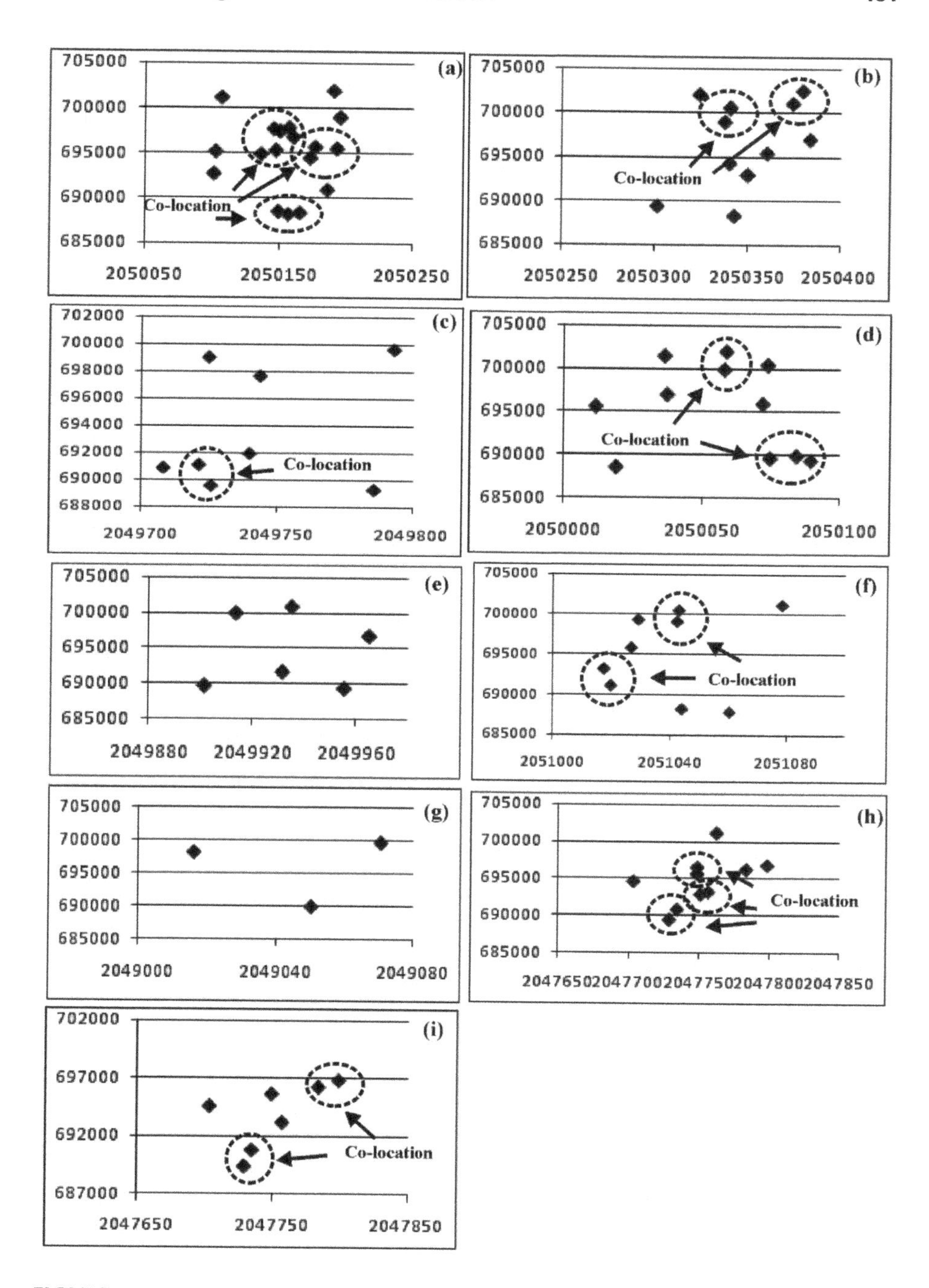

FIGURE 7.10 Spatial distributions of SKP instances after initial determination of co-location algorithm.

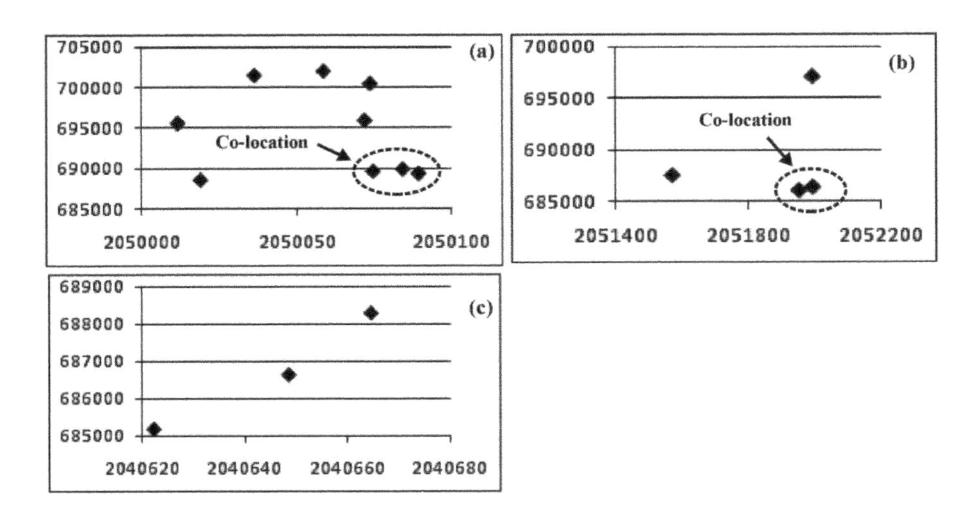

FIGURE 7.11 Spatial distributions of SO instances after initial determination of co-location algorithm.

instances. The computation of cross-correlation is modeled in equation 7.8, with which we have a co-correlation matrix, as follows:

$$\underset{946\times946}{\Sigma} = \begin{bmatrix} \rho_{11} & \rho_{12} & \cdots\cdots & \rho_{1\times946} \\ & \rho_{22} & \cdots\cdots & \rho_{2\times946} \\ & & \cdots\cdots & \cdots \\ & & \cdots & \rho_{945\times946} \\ & & & \rho_{946\times946} \end{bmatrix} \tag{7.8}$$

With observation to the coefficient of cross-correlation matrix, cross-correlation coefficients threshold is set at 0.95, that is,

$$p_{i,j} = \begin{cases} \text{non-correlated} & \text{if } p_{i,j} \leq 0.95 \\ \text{correlated} & \text{if } d_{i,j} > 0.95 \end{cases} \qquad d_{i,j} \subseteq \sum_{946\times946} \tag{7.9}$$

With the given threshold, all of the candidates' prevalent co-location events are kept without pruning.

7.3.3 Generating Co-Location Rules

Accompanying with the generation of co-location set, the co-location rules with the user-defined constrain conditions (threshold) from the prevalent co-locations and their table instances can be generated (see Figure 7.12), that is,

IF (Dist_ij <= 25000 **AND** Gramma <=85 **AND** cross-correlation<=0.95) Co-location
ELSE Non_Co-location

FIGURE 7.12 Generated co-location rules.

7.4 EXPERIMENTS OF CL-DT INDUCTION

7.4.1 Basic Steps of CL-DT Induction

With the previously generated prevalent co-location events, the CTree for Excel tool is applied to create the decision tree. The steps include (Zhou and Wang 2010).

7.4.1.1 Step 1. Load Pavement Database

As described in sections 7.2 and 7.3, the pavement database first has to be loaded into the software. In this experiment, the loaded data are a prevalent co-location database, that is, they have been "preprocessed" using a co-location mining algorithm. Similarly, the distress data (nonspatial data) in the database, such as N, L, M, and S, will be quantified into 100, 75, 50, and 25.

7.4.1.2 Step 2. Data Inputs

Similarly, some parameters to optimize the processes of decision tree generation will be input. These parameters include:

1. *Adjust factor of categorical predictor*: While growing the tree, child nodes are created by splitting parent nodes; which one is a predictor to use for this split is decided by a certain criterion. Because this criterion has an inherent bias towards choosing predictors with more categories, thus, the input of an adjust factor will be able to adjust this bias.
2. *Minimum node size criterion*: While growing the tree, whether to stop splitting a node and declare the node as a leaf node will be determined by some criteria that we need to choose. These criteria are the same as those adopted in section 7.4.1, i.e.:
   a. *Minimum node size*: A valid minimum node size is between 0 and 100.
   b. *Maximum purity*: An effective value is between 0 and 100. Stop splitting a node if its purity is 95% or more. Also, stop splitting a node if the number of records in that node is 1% or less of the total number of records.
   c. *Maximum depth:* a valid maximum depth is greater than 1 and less than 20. Stop splitting a node if its depth is 6 or more.
3. *Pruning option:* This option allows us to decide whether or not to prune the tree when the tree is growing, which can help us to study the effect of pruning.
4. *Training and test data:* In this research, a subset of data is used to build the model and the rest to study the performance of the model. Also, a random selection of the test set at a ratio of 10% is adopted.

TABLE 7.7
Information of the Induced Decision Tree Using CL-DT Algorithm

Tree Information		% Misclassified		Time Taken (Second)	
Total Number of Nodes	22	Training Data	21.7%	Data Processing	1
Number of Leaf Nodes	14	Test Data	15.3%	Tree Growing	2
Number of Levels	13			Tree Pruning	1
				Tree Drawing	5
				Classification using final tree	1
				Rule Generation	19

7.4.2 Experimental Results

7.4.2.1 Induced Decision Tree

With the previous data input, a decision tree is generated. The corresponding information for the decision tree, including misclassified data percentage, time taken, total number of nodes, number of leaf nodes, and number of levels is listed in Table 7.7.

7.4.2.2 Induced Decision Rules

After the decision tree is induced, the tree is further processed to induce decision rules. The decision rules are directly induced in this research by forming a conjunct of every test that occurs on a path between the root node and a leaf node of a tree, that is, top-to-bottom mode. Thus, the decision rules are first induced by ordering all the classifications and then using a fixed sequence to combine them together. After this processing, 12 rules are generated. Finally, seven rules are induced and depicted in Figure 7.13. The quality of the individual rules is measured by Support, Confidence, and Capture (see Table 7.8).

TABLE 7.8
Support, Confidence, and Capture for Each Generated Rule

Rule ID	Classes	Support	Confidence	Capture
0	NO	100.0%	89.2%	92.2%
1	CP	80.0%	95.0%	79.9%
2	SKP	79.2%	83.6%	83.8%
3	FDP	83.2%	84.4%	77.7%
4	PM2	83.5%	94.1%	79.2%
5	PM1	89.0%	88.8%	90.2%
6	SO	83.3%	76.3%	89.5%

Rule 1:
IF (AN" >=5 AND "AS_" =0 AND "BK" ='50' AND "RF" ='100' AND "RT"='75' OR "RT"='100' AND "RV"='100' AND "RQ"='75' AND "RATING" > '73')
THEN CP

Rule 2:
IF ("AN" >=3 AND "AS_" >=1 AND "BK" ='100' AND "RF" ='100' AND "RV"='100' OR "RV"='75' AND "RQ"='75' OR "RQ"='50' AND "RATING" >= '43')
THEN FDP

Rule 3:
IF ("AN" >=2 AND "AN" <= 6 AND "AS_" >=0 AND "BK" ='100' AND "RF" ='100' AND "RV"='100' AND "RQ"='75' AND "RATING" >= '45' AND "RATING" <= '65').
THEN PM1

Rule 4:
IF ("AN" >=2 AND "AN" <= 6 AND "AS_" >=0 AND "BK" ='100' AND "RF" ='100' AND "RV"='100' AND "RQ"='75' AND "RATING" >= '55' AND "RATING" <= '70')
THEN PM2

Rule 5:
IF "AN" >3 AND "AM" >1 AND "AS_" =0 AND "RF"='100' AND "RV"='100' OR "RV"='75' AND "RQ"='75' AND "RATING" >= '63' AND "RATING" < '90')
THEN SKP

Rule 6:
IF "AN" >=6 AND "AM" =0 AND "AS_" =0 AND "BK"='100' AND "RF"='100' AND "RV"='100' OR "RV"='75' AND "RQ"='75' AND "RATING" >= '60' AND "RATING" <= '75')
THEN SO

Rule 7:
IF "AN" >=9 AND "RATING" <= '100')
THEN Nothing

FIGURE 7.13 The final rules after verification and post-processing.

7.5 MAPPING OF CL-DT-BASED DECISION OF M&R

With these rules induced, the M&R strategies can be predicted and decided for each road segment in the database using the rules. In other words, the operation using the co-location decision tree only occurs in the database, and thus the results cannot be visualized and displayed on either map or screen. Thus, this research employed ArcGIS software in combination with the above induced results to create the map of decision making for maintenance and rehabilitation. The basic operation is the same as that described in Chapter 7.5, that is, taking each rule as a logic query in ArcGIS software, then queried results are displayed in the ArcGIS layout map. In order to compare the results, the rehabilitations suggested by engineers at the ITRE of North Carolina State University are superimposed with the decisions made at this research. As seen from Figure 7.11 through Figure 7.19, each rehabilitation strategy derived in this research can be located with its geographical coordinates and visualized with its spatial data, nonspatial data, and different gray.

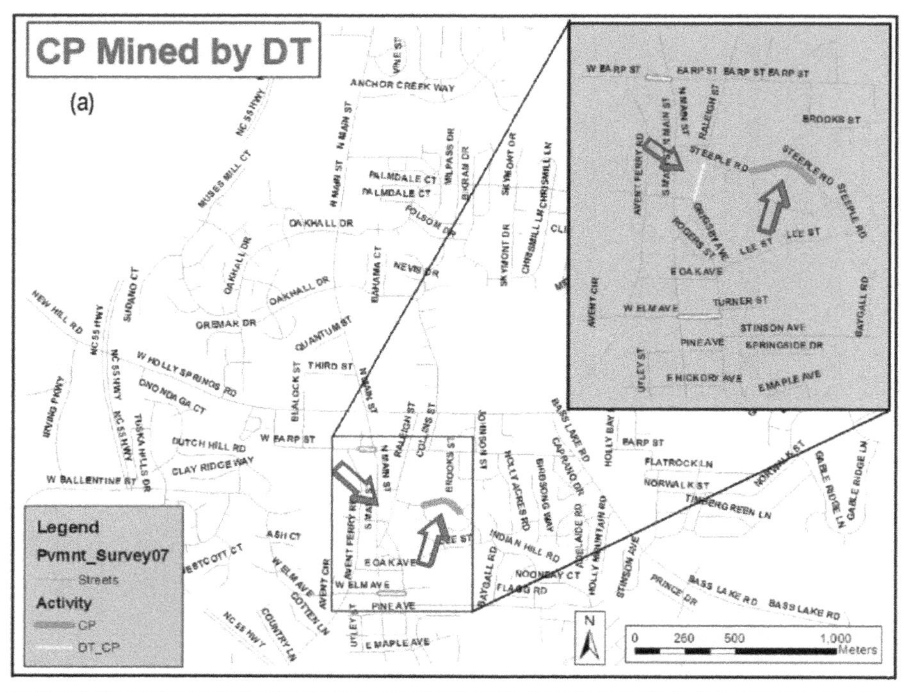

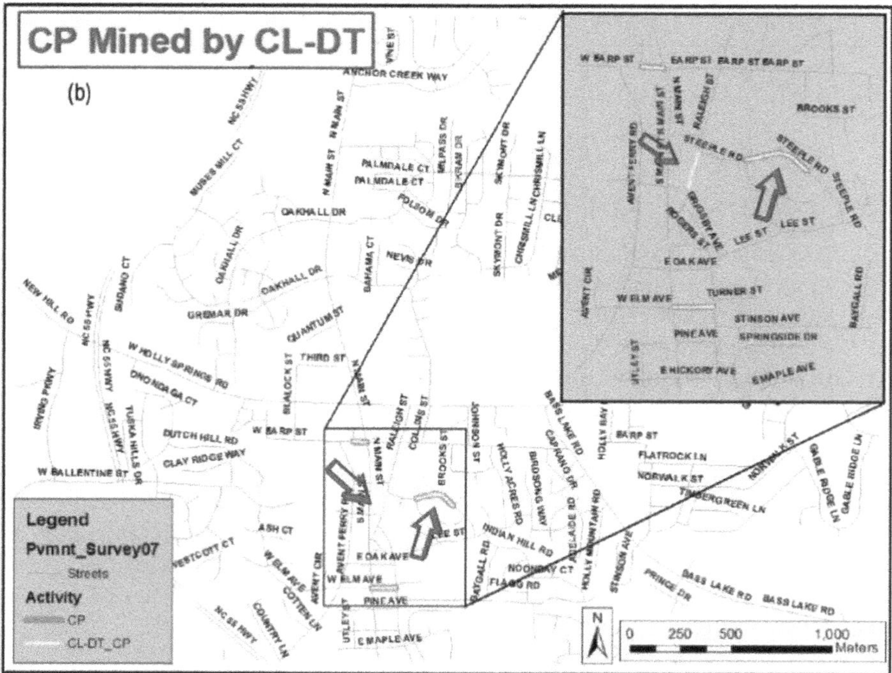

FIGURE 7.14 Comparison analysis of the CP decision of road rehabilitation made by DT (described in section 7.3) and the proposed CL-DT (described in section 7.5), both of which are compared to the CP decision made (provided) by the ITRC at the NCDOT.

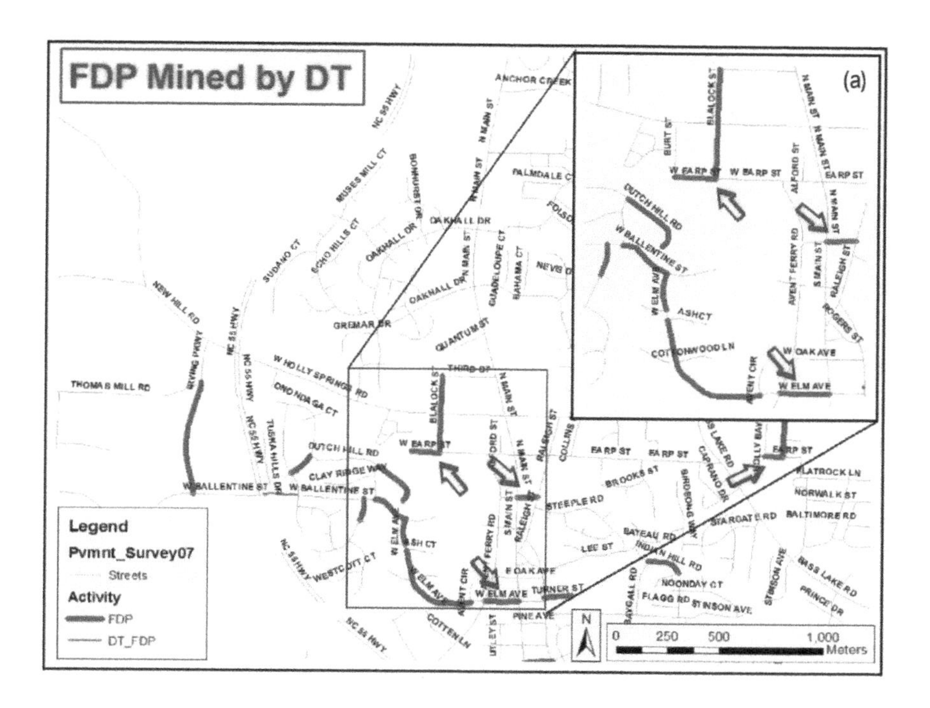

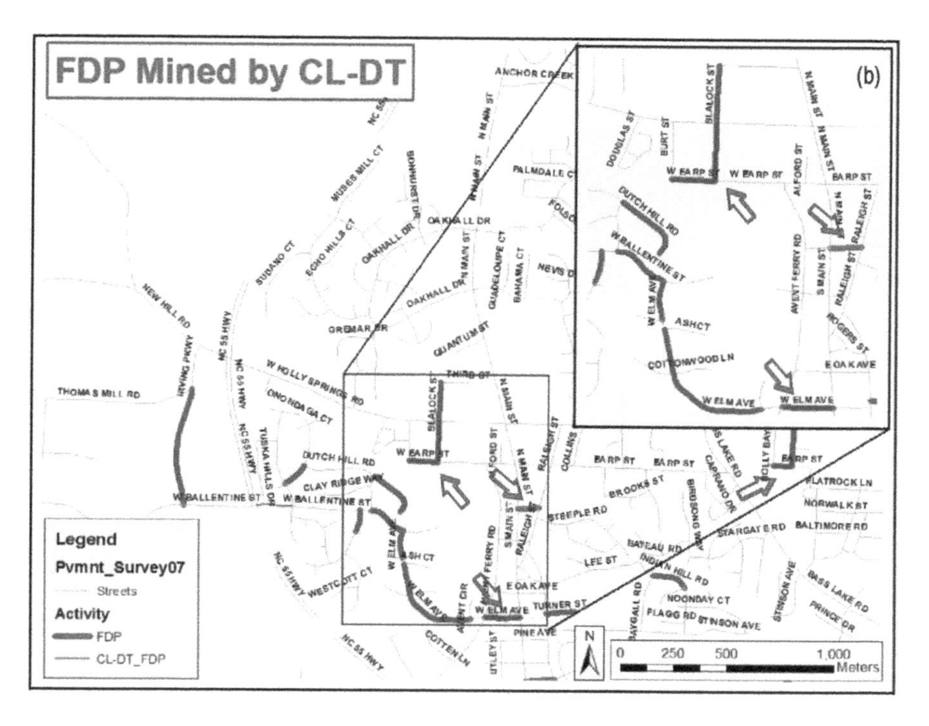

FIGURE 7.15 Comparison analysis of the FDP decision of road rehabilitation made by DT (described in section 7.3) and the proposed CL-DT (described in section 7.5), both of which are compared to the FDP decision made (provided) by the ITRC at the NCDOT.

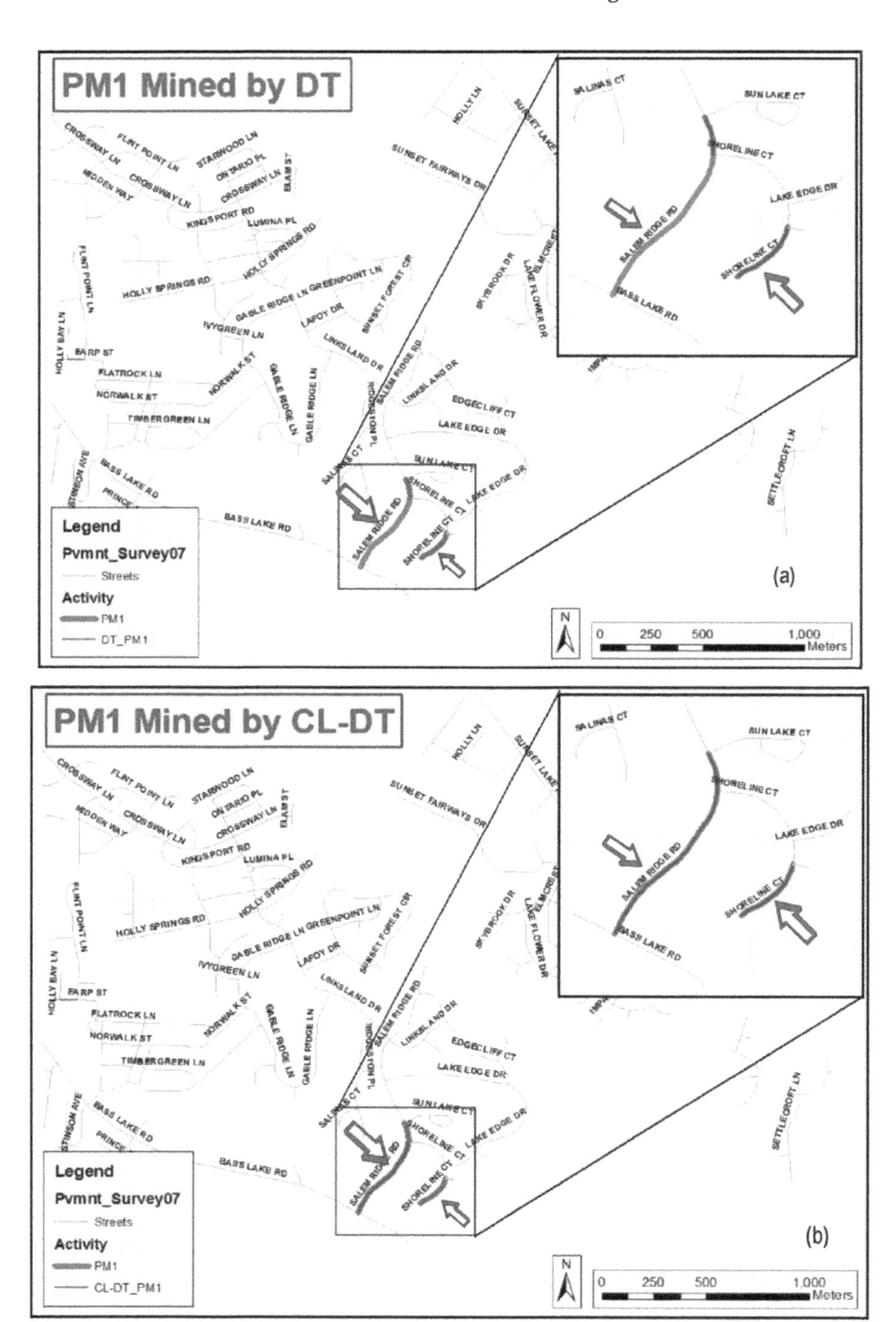

FIGURE 7.16 Comparison analysis of the PM1 decision of road rehabilitation made by DT (described in section 7.3) and the proposed CL-DT (described in section 7.5), both of which are compared to the PM1 decision made (provided) by the ITRC at the NCDOT.

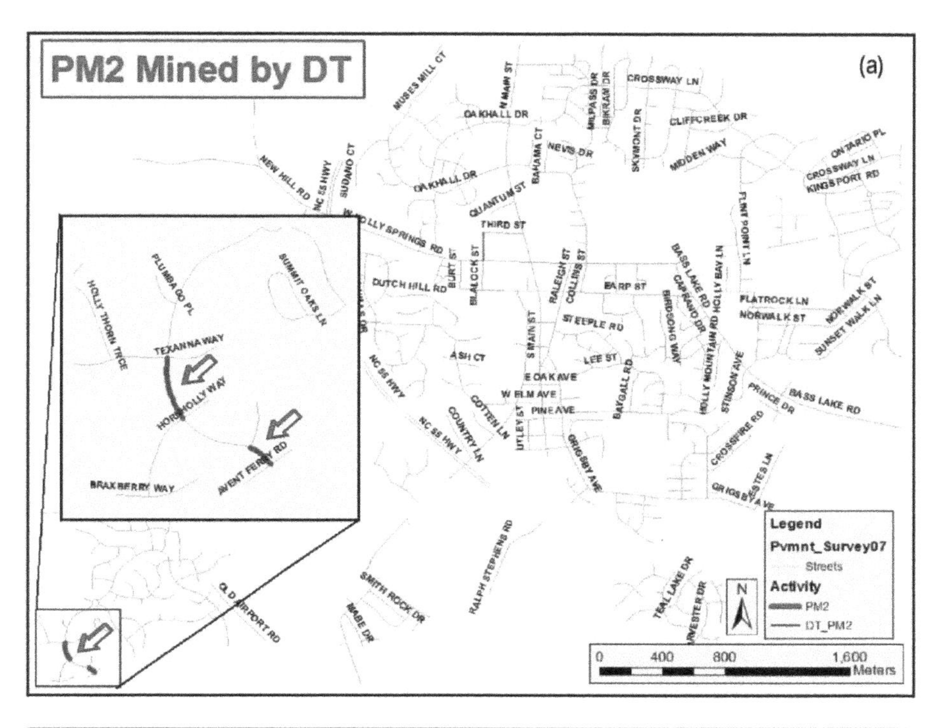

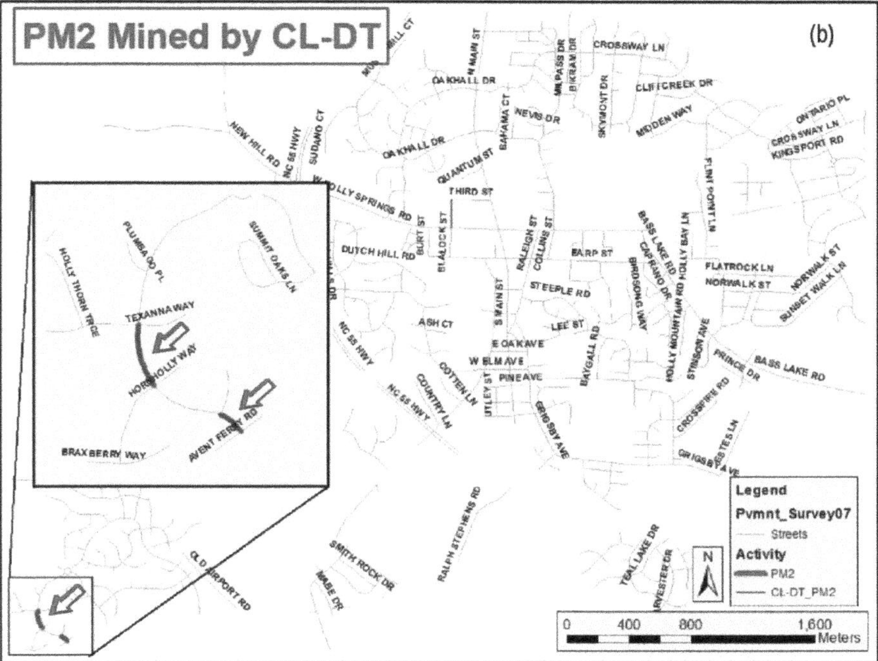

FIGURE 7.17 Comparison analysis of the PM2 decision of road rehabilitation made by DT (described in section 7.3) and the proposed CL-DT (described in section 7.5), both of which are compared to the PM2 decision made (provided) by the ITRC at the NCDOT.

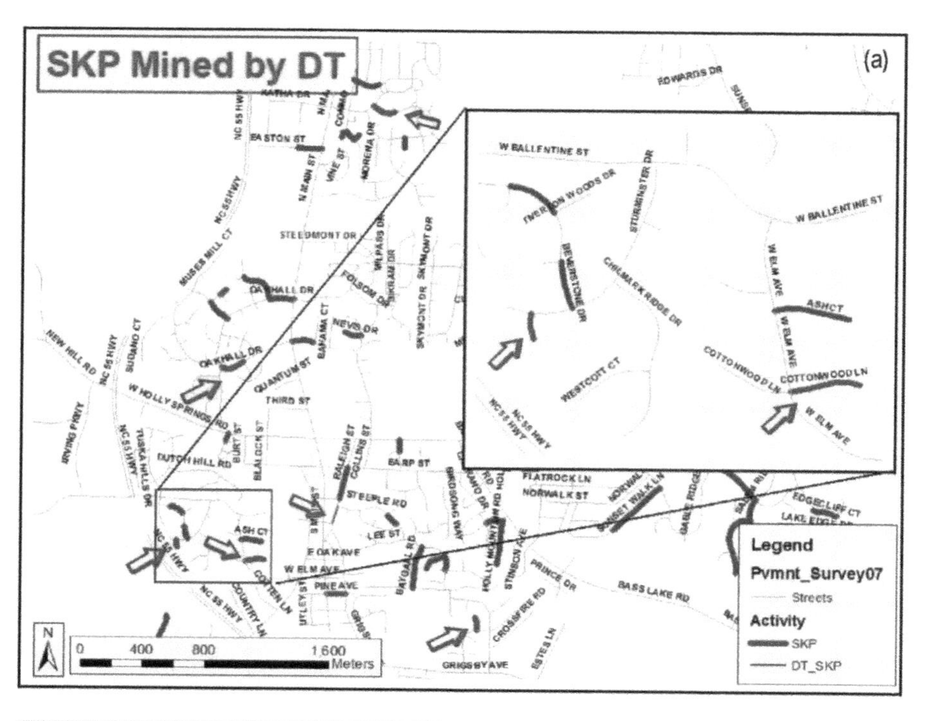

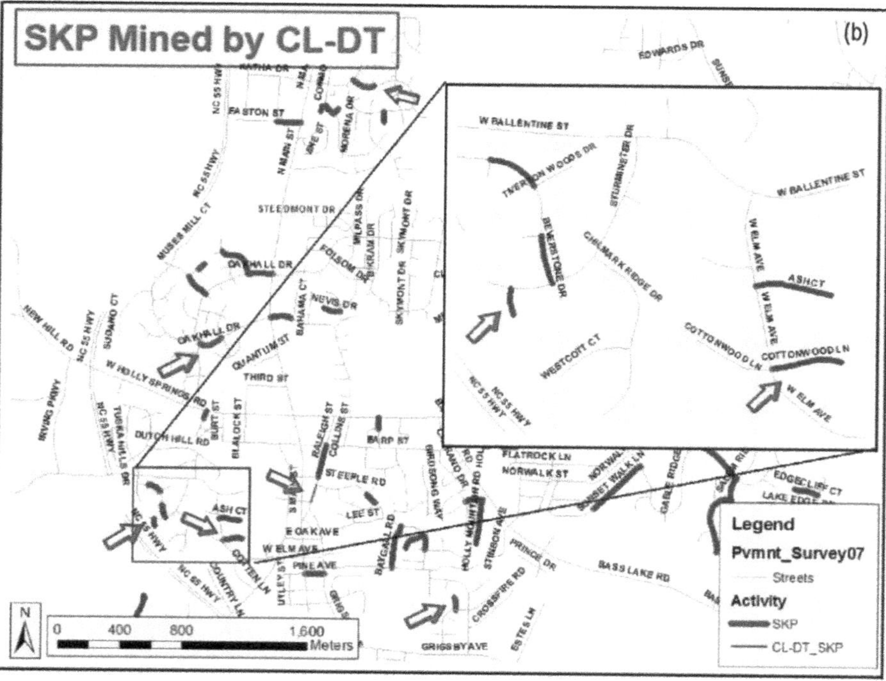

FIGURE 7.18 Comparison analysis of the SKP decision of road rehabilitation made by DT (described in section 7.3) and the proposed CL-DT (described in section 7.5), both of which are compared to the SKP decision made (provided) by the ITRC at the NCDOT.

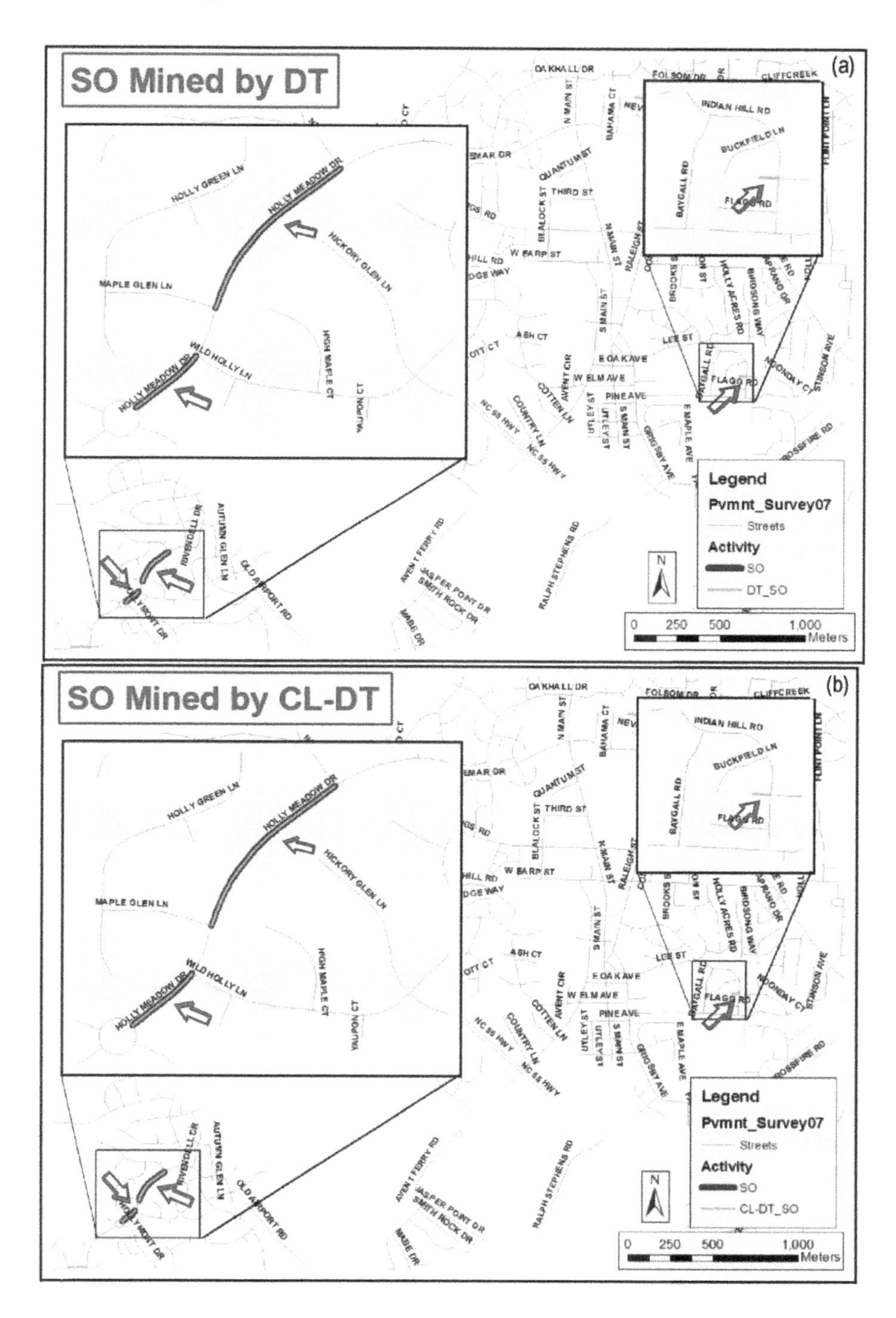

FIGURE 7.19 Comparison analysis of the SO decision of road rehabilitation made by DT (described in section 7.3) and the proposed CL-DT (described in section 7.5), both of which are compared to the SO decision made (provided) by the ITRC at the NCDOT.

7.6 COMPARISON ANALYSIS AND DISCUSSION

7.6.1 Comparison Analysis for the Induced Decision Tree Parameter

The proposed co-location decision tree (CL-DT) method should have many advantages over the traditional decision tree method in the effectiveness and accuracy of decision tree (decision rules) generation when applied in the decision making of road maintenance and repair. In order to validate this conclusion, we compare the tree induction information for the two methods, and the results are listed in Table 7.9. As seen in Table 7.9, the total number of nodes, number of leaf nodes, and number of levels decreases 51%, 62%, and 35%, respectively. Thus, computational time will largely decrease. Accuracy of the decision tree increases.

7.6.2 Comparison Analysis for the Misclassified Percentage

Also, we check the misclassified percentage, and the results are listed in Table 7.10. As seen in Table 7.10, the misclassified percentage for the training data decreases from 61.2% to 9.7%. This is probably caused by the fact that we used co-location mining technology to delete any non-prevalent candidate co-location instances. As a result, the training data contributed to the decision tree induction.

TABLE 7.9
Comparison of Tree Information Parameters between DT and CL-DT Algorithm

Tree Information	Methods		Decreasing percentage
	DT	CL-DT	
Total Number of Nodes	72	35	51%
Number of Leaf Nodes	37	14	62%
Number of Levels	20	13	35%

TABLE 7.10
Comparison of Misclassified Percentage between DT and CL-DT Algorithm

Misclassified percentage	Methods	
	DT	CL-DT
Training Data	61.2%	9.7%
Test Data	60.0%	8.3%

TABLE 7.11
Comparison of the Computation Time between DT and CL-DT Algorithm

Items	Time taken for two methods (second)		% decreasing
	DT	CL-DT	
Data Processing	1	1	Rounded to 1"
Tree Growing	6	2	66%
Tree Pruning	1	1	Rounded to 1"
Tree Drawing	10	4	60%
Classification using final tree	1	1	Rounded to 1"
Rule Generation	35	15	20%

7.6.3 Comparison Analysis for the Computational Time

Theoretically, the proposed CL-DT method should save much computational time, since the "preprocessing" method uses co-location mining technology, which deletes the non-prevalent co-location events. In order to verify this conclusion, we retrieved the computational time of data processing, tree growing, tree pruning, tree drawing, classification using final tree, and rule generation from the computer for the two methods. The results are listed in Table 7.11. As observed in Table 7.11, the time taken for the tree growing, tree drawing, and rule generation is largely decreased. The time taken for rule generation decreases by 20%.

7.6.4 Comparison Analysis of Support, Confidence, and Capture for Rule Induction

Another comparison analysis is for support, confidence, and capture of training data when inducing the decision rules. The results for the two methods are listed in Table 7.12. As observed in Table 7.12, the percentage of support, confidence, and capture for training data in FDP treatment strategy increase from 71.6%, 55.6%, and 66.2% to 83.2%, 84.4%, and 77.7%, respectively. This means that most of training data actively contributes the decision rule induction, which demonstrates that the co-location mining method can largely increase the effectiveness of decision tree/rules induction.

7.6.5 Verification of the Quantity of Each Treatment Strategy

As mentioned earlier, the ITRC at the NCDOT has indicated the quantity of six treatment strategies at different road segments in the study area (four counties at North Carolina). Theoretically, the proposed CL-DT method should find the same quantity and location of each treatment strategy as those proposed by the ITRC at the NCDOT, since the proposed CL-DT applied the experts' knowledge from the

TABLE 7.12
Comparison for Support, Confidence and Capture for Two Methods in Each Generated Rule

Rule ID	Strategies	Support		Confidence		Capture	
		DT	CL-DT	DT	CL-DT	DT	CL-DT
0	NO	100.0%	100.0%	86.7%	89.2%	93.0%	92.2%
1	CP	60.7%	80.0%	100.0%	95.0%	75.6%	79.9%
2	SKP	60.5%	79.2%	66.7%	83.6%	82.5%	83.8%
3	FDP	71.6%	83.2%	55.6%	84.4%	66.2%	77.7%
4	PM2	80.2%	83.5%	100.0%	94.1%	73.1%	79.2%
5	PM1	81.3%	89.0%	71.4%	88.8%	85.3%	90.2%
6	SO	81.6%	83.3%	66.7%	76.3%	73.5%	89.5%

TABLE 7.13
Quantity Comparison of Different Treatment Strategies Made by Three Methods

ID	Proposed treatment strategies	Methods	Quantity	Differences in quantity referred to NCDOT
1	Crack Pouring (CP)	NCDOT	3	
		DT	3	0
		CL-DT	3	0
2	Full-Depth Patch (FDP)	NCDOT	34	
		DT	29	5
		CL-DT	32	2
3	1" Plant Mix Resurfacing (PM1)	NCDOT	6	
		DT	7	1
		CL-DT	6	0
4	2" Plant Mix Resurfacing (PM2)	NCDOT	3	
		DT	4	1
		CL-DT	5	2
5	Skin Patch (SKP)	NCDOT	65	
		DT	56	9
		CL-DT	62	3
6	Short Overlay (SO)	NCDOT	3	
		DT	5	2
		CL-DT	4	1

ITRC. In order to verify this result, Table 7.13 lists the comparison for each treatment strategy proposed by ITRC at the NCDOT and discovered by the proposed CL-DT method. Meanwhile, the quantity of each treatment strategy discovered by the DL method (Zhou and Wang 2010) is also listed in Table 7.13.

As observed in Table 7.13, the quantity discovered by CL-DT is very close to those proposed by the ITRC for each treatment strategy. Thus, the traditional decision tree method mines 56 skin patch (SKP) strategies, which is nine differences from those proposed by the ITRC, while the CL-DT method mined 62 SKP treatments, which is only three differences from those proposed by the ITRC.

7.6.6 Verification of the Location of Each Treatment Strategy

Also, the ITRC at the NCDOT has indicated locations of six treatment strategies at different road segments in the study area (four counties at North Carolina). Theoretically, the proposed CL-DT method should find the same location for each treatment strategy as those proposed by the ITRC at the NCDOT, since the proposed CL-DT applied the experts' knowledge (distress for each road segment) from the ITRC. In order to verify this conclusion, Table 7.14 lists the comparison for each treatment strategy proposed by ITRC at the NCDOT and discovered by the proposed CL-DT method. Meanwhile, the locations of each treatment strategy discovered by DL method (Zhou and Wand 2010) are also listed in Table 7.14.

As observed in Table 7.14, the location differences referred to those proposed by CL-DT for skin patch (SKP) strategies is significant. In other words, 13 road segments for SKP strategy are different from those proposed by the traditional decision

TABLE 7.14
Location Comparison of Different Treatment Strategies Made by Three Methods

ID	Proposed treatment strategies	From	Number	Difference in location referred to NDCOT
1	Crack Pouring (CP)	NCDOT	3	
		DT	3	1
		CL-DT	3	1
2	Full-Depth Patch (FDP)	NCDOT	34	
		DT	29	3
		CL-DT	32	1
3	1" Plant Mix Resurfacing (PM1)	NCDOT	6	
		DT	7	1
		CL-DT	6	0
4	2" Plant Mix Resurfacing (PM2)	NCDOT	3	
		DT	4	1
		CL-DT	5	1
5	Skin Patch (SKP)	NCDOT	65	
		DT	56	13
		CL-DT	62	3
6	Short Overlay (SO)	NCDOT	3	
		DT	5	2
		CL-DT	4	1

tree method, but only three differences by the CL-DT method, when referred to those by the ITRC (also see Figure 7.9a and 7.9b).

7.7 DISCUSSION AND REMARKS FOR CO-LOCATION DECISION TREE ALGORITHMS

With the existing shortcomings of the decision tree induction method discovered in Chapter 1.5, this chapter presented the theory and algorithm of a new decision tree induction, called a *co-location decision tree (CL-DT)*. The main purpose of the proposed algorithm is to utilize the characteristics of attribute co-location (co-occurrence) to find the co-occurrence rules. These rules are used to enhance the traditional decision tree induction algorithm.

With the described experimental results and comparison analysis, it can be concluded that the proposed CL-DT algorithm can better make a decision for pavement treatment maintenance and rehabilitation when compared to the traditional decision tree method (e.g., C5.0 algorithm), since the new proposed method considers the co-occurrence distinct events. This chapter especially makes a comparison analysis for the induced decision tree parameter, the misclassified percentage, the computational time taken, support, confidence and capture for rule induction. This chapter also verified the quantity and location of each treatment strategy referred to those proposed by the ITRC at the NCDOT.

With these experimental results and comparison analyses, it can be concluded that:

1. The proposed CL-DT method has many advantages over the traditional decision tree method in the effectiveness and accuracy of decision tree (decision rules) generation when applied in the decision making of road maintenance and repair. With comparing the analyses of two methods, DT and CL-DT, it is concluded that the total number of nodes, number of leaf nodes, and number of levels decrease 51%, 62% and 35%, respectively.
2. With comparison analysis of two methods, DT and CL-DT, it is concluded that the misclassified percentage for the training data decrease from 61.2% to 9.7%, which demonstrated that the training data can fully play roles in contribution to decision tree induction.
3. With the comparison of the two methods, DT and CL-DT, it is concluded that the time taken by data processing, tree growing, tree pruning, tree drawing, classification using final tree, and rule generation is largely decreased, which can achieve 20% for rule generation.
4. With the comparison of the two methods, it is concluded that the percentage of support, confidence, and capture for the FDP treatment strategy increase from 71.6%, 55.6%, and 66.2% to 83.2%, 84.4%, and 77.7%, respectively. This means that most of the training data contributes to the decision rule induction.
5. With comparison of the quantity of six treatment strategies proposed by the ITRC at different road segments in the study area and by the CL-DT method, it is concluded that that the quantity discovered by CL-DT is much closer to those proposed by the ITRC for each treatment strategy. For example, 56 skin

patch (SKP) strategies were mined by the traditional decision tree method, which is nine differences from those proposed by the ITRC, while there were only three differences for the proposed CL-DT method when compared to those proposed by the ITRC.

With comparison of the locations of six treatment strategies at different road segments in the study area proposed by CL-DT method and by the ITRC at the NCDOT, it is found that there are 13 road segments for SKP strategy different from those proposed by the traditional decision tree method but only three differences from the CL-DT method when compared to those by the ITRC.

7.8 CONCLUSIONS

This chapter verified the development of the theory and algorithm of a new decision tree induction algorithm called *co-location-based decision tree (CL-DT)*.

Through this experiment analysis, the following findings have been discovered and the following conclusions have been drawn up.

7.8.1 Advantages and Disadvantages of Applying Existing DT Method in Pavement M&R Strategy Decision Making

The advantages of applying the existing DT method are (1) the DT technology can make the consistent decision of pavement M&R strategy under the same road conditions, that is, with less interference from human factors. (2) The DT technology can largely increase the speed of decision making because the technology automatically generates decision tree and decision rules if the expert knowledge is given and thus, largely saves time and cost of pavement management. (3) Integration of the DT and GIS can provide the PMS with the capabilities of graphically displaying treatment decisions, visualize the attribute and non-attribute data, and link data and information to the geographical coordinates.

Disadvantages of applying the existing DT method are (1) existing DT induction methods are not as quite intelligent as people's expectation. In other words, the DT inducted by DMKD are not completely exact; thus, post-processing and refinement are necessary. (2) Existing DT induction methods for pavement M&R strategy decision making only used the nonspatial attribute data. It has been demonstrated that the spatial data is very useful for enhancing decision making of pavement treatment strategies. (3) A DT induction method is based on the knowledge acquired from pavement management engineer for strategy selection. A decision tree is in fact to organize the obtained knowledge in a logical order. Thus, the decision trees can determine the technically feasible rehabilitation strategies for each road segment.

7.8.2 Significances of the Proposed CL-DT Method for Pavement M&R Strategy Decision Making

This research has verified the advantages through experimental results and several comparison analyses including the induced decision tree parameters; the misclassified percentage; the computational time taken, support, confidence, and capture for rule induction; and the quantity and location of each treatment strategy. It can be concluded that

a. The proposed CL-DT algorithm can better make a decision for pavement M&R strategy when compared to the existing decision tree method (e.g., C4.5 algorithm), since the new proposed method considers the co-location (co-occurrence) distinct events of spatial data in the pavement database.
b. The proposed CL-DT method has higher accuracy and effectiveness than the existing decision tree method does. The induced tree information, including the total number of nodes, number of leaf, and number of levels, decrease 51%, 62%, and 35%, respectively.
c. The training data can fully play a role in contribution to decision tree induction. For example, the misclassified percentage for the training data using the CL-DT method decreased from 61.2% to 9.7%; the percentages of support, confidence, and capture of the FDP treatment strategy increased from 71.6%, 55.6%, and 66.2% to 83.2%, 84.4%, and 77.7%, respectively.
d. The computational time taken for the tree growing, tree drawing, and rule generation is largely decreased for CL-DT method, which achieved 20% for the rule generation.
e. The quantity of six treatment strategies proposed by the ITRC and by CL-DT method at different road segments in the study area is much closer for each treatment strategy.
f. Locations of six treatment strategies proposed by the CL-DT method and by the ITRC at different road segments in the study area are close, but different for the existing DT method.

REFERENCES

Greene, J., and Shahin, M., Airfield pavement condition assessment. Washington, DC: *U.S. Army Corps of Engineers*, September 2010.

Zhou, G., Co-location decision tree for enhancing decision-making of pavement maintenance and rehabilitation. *Ph.D. dissertation*, Virginia Tech, Blacksburg, Virginia, USA, 2011.

Zhou, G., and Wang, L., GIS and data mining to enhance pavement rehabilitation decision-making. *Journal of Transportation Engineering*, vol. 136, no. 4, February 2010, pp. 332–341.

Zhou, G., and Wang, L., Co-location decision tree for enhancing decision-making of pavement maintenance and rehabilitation. *Transportation Research Part C*, vol. 21, 2012, pp. 287–305.

8 Application of Mining Co-Location Patterns in Buffer Analysis

8.1 INTRODUCTION

The research for buffering algorithms can traced back to the 1960s (Shimrat 1962), when Europe first introduced statistics into geography. The traditional buffering algorithm includes point elements, line elements, and polygon element buffering. Since the 1960s, many novel algorithms have been developed. For example, in early 1975, Chvátal (1975) first proposed a buffer creation algorithm using computational geometry, and then Preparata and Shamos (1985) and Aggarwal et al. (1985) extended this method. Wu (1997) proposed an improved algorithm for generation of line buffering zone called a "geometric model for both-side parallel lines buffer generation." This method was presented on the basis of a comparison analysis of the two methods of line buffer generations, the angle bisector method and the circular arc method, and can effectively solve problems such as determination of convex and concave of a chord arc. Wu (Wu et al. 1999) proposed a vector buffer algorithm for point, line, and polygon elements using a buffer curve and edge-constrained triangulation network. This method can efficiently reduce the computational complication in the process of cutting and reorganizing the buffer boundary line or/and curve. Chen et al. (2004) proposed a method called Voronoi *k*-order neighbors to recreate the buffer area. Dong et al. (2003) improved the buffer generation algorithm with double parallel lines and arcs using the rotation transformation and the recursive. This method can largely simplify the process of line buffer generation and meanwhile efficiently corrects the acute corners of the boundary of the area buffer. This method was considered the better solution for the intersection of line buffer generation. In order to reduce the time consumed during generating the point buffering in the self-intersection section, Ren et al. (2004) proposed the application of the Douglas-Peuker algorithm to extract the feature points of the curve before the buffer was generated. The experimental result demonstrated that this algorithm can largely decreased computational time. Er et al. (2009) proposed a novel method in which the buffer distance is variable during generating line buffers. However, the variable buffer distances are indeed equidistant, and as a result, this method has met challenges for complicated and irregular regions with various geographic attributes. Pan and Li (2010) proposed a random algorithm for buffer zone generation. This algorithm can be carried out using computer parallel processing and consequently largely improves the efficiency of buffer zone generation.(Liu and Min (2011) conducted a comprehensive overview for various buffer algorithms, analyzed and compared the advantages and disadvantages of these early methods, and concluded the most appropriate solution to the problems such as distortion and correction

DOI: 10.1201/9781003139416-8

of the acute corners during the process of buffer generation. Bai and Zhang (2011) proposed the convex arc algorithm for solution of the problems of self-intersection and acute corner correction. This method can effectively solve the problem of area buffer generation with multiple embedded inner rings. Zhang et al. (2014) propose a buffering approach for matching areal objects (e.g., buildings) on the basis of relaxation labeling techniques. This method has successfully been applied in pattern recognition and computer vision. Chen and Liu (2014) proposed the function form of the buffer expression varying with a certain variable, which effectively solved the problem of determining the complex boundary of the buffer on both sides of the convex and concave inflection points under the variable buffer distance. Xu and Liu (2014) proposed a vector grid hybrid algorithm to solve the problem of buffer generation for line targets. First, convert the vector data to a raster format, second, use the Douglas-Peuker algorithm to resample the line, then generate a buffer based on the expansion principle, and finally deal with the problem of buffer self-intersection. Yang et al. (2014) used GPU (Graphic Processor Unit) to calculate the distance between vector data, which can quickly realize the cache analysis of vector data and also solve the problem of buffer boundary self-intersection. This method has successfully been applied in three-dimensional (3D) digital earth. Fan et al. (2014) proposed "a parallel buffer algorithm to improve the performances of buffer analyses on processing large datasets," which was based on area merging and massage passing interface. But in the algorithm, the relationship of adjacent vector features was not considered. Wang E. et al. (2016) proposed a new method that first classifies attribute data and then establishes multiple temporal buffers, finally smoothing the boundary line using the Douglas-Peuker algorithm and Bézier curve. Wang T. et al. (2016) and Huang (2020) proposed a parallel algorithm on the basis of the equal arc segmentation method, which first segments a whole arc into many segments, then takes the segments as a unit, and finally deals with each of the units, respectively. Dong and Jing (2017) proposed buffer generation method on the basis of a geodesic of tiles. The buffer distances on both sides are different and are measured using geodetic facets. Chen (2018) present a buffering method using the probability and entropy theory. This method can effectively reduce the error of spatial analysis and avoid the error of spatial decisions. Lee and Who (2011) used a diagram of long distance and polar angle bins to compute the correlation. Ma et al. (2019) proposed a buffer generation method on the basis of a combination of spatial index and data transformation in order to speed up data processing.

Although many efforts have been made over 60 years, the traditional buffering methods have encountered many challenges in practice due to the increasing requirement in buffering zone accuracy. This is because the existing buffering methods are based on a fixed buffer distance without considering the differences of attributes within the buffer zone, that is, the existing methods treat all attributes within buffer zone as the same, that is, homogenous. As a result, the resulting buffer zone deviates from the "true" buffer zone and loses accuracy. For example, as shown in Figure 8.1, given point P and white star targets (such as A_0, A_1, etc.) are homogeneous patterns. The boundary of traditional buffer zone of point P is the arc of $A_0BA_4A_3A_0$, of which the buffer radius is a fixed value r. However, the attributes of black star target B in traditional point buffer zone are heterogeneous with the other targets. Additionally, the white star A_1 outside the buffer zone is homogeneous with the others. Therefore,

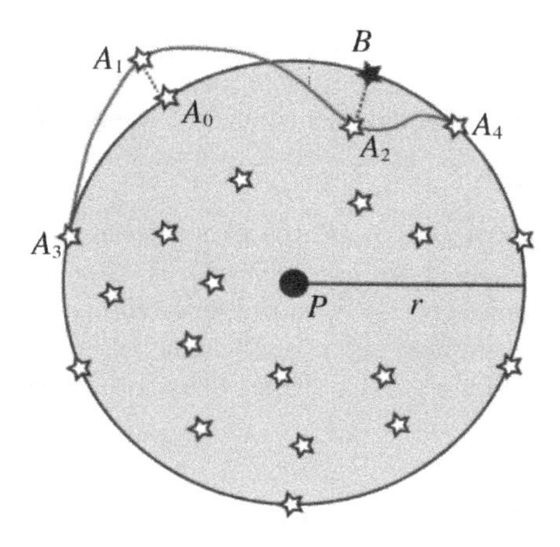

FIGURE 8.1 Traditional point buffer zone (surrounded by the arc of $A_0BA_4A_3A_0$) and generalized point buffer zone (surrounded by the arc of $A_1A_2A_4A_3A_1$).

the buffer zone of point P should NOT be a regular shape, while the irregular shape, that is, the zone surrounded by the arc of $A_1A_2A_4A_3A_1$, is more in line with the actual situation. For this reason, a radical algorithm called a Generalized buffering algorithm (GBA) is proposed for buffer zone generation.

8.2 GENERALIZED BUFFERING ALGORITHMS

From a viewpoint of spatial entities, the generation of a buffer zone is to build a polygon around the point, line, and polygon instances. On the other hand, from a viewpoint of mathematics, the generation of a buffer zone is to determine the neighborhood of one certain spatial object and the range of a neighborhood upon the radius of neighborhood (called *buffer radius*, or *buffer distance* in a traditional buffering algorithm). At this point, the traditional buffering algorithm can be described as follows:

Let $A = \{A_1, A_2, \ldots, A_d\}$ be a set of objects, and then the buffering of one spatial object is expressed by

$$Buffer_j = \left\{ p : d(p, A_j) \leq r \right\} \tag{8.1}$$

where r, which is a constant, is the buffer radius of the buffer zone, and $d(p, A_j)$ is the Euclidean distance between p and A_j. $Buffer_j$ is the instances set where the distance is less than or equal to r.

For the set of spatial objects $A=\{A_1, A_2, \ldots, A_d\}$, the buffer zone can be defined by

$$Buffer = \bigcup_{j=1}^{d} Buffer_j \tag{8.2}$$

As mentioned in section 8.2, the existing buffering algorithms have shortcomings, such as the equidistance of buffer distance and ignoring the homogenous and/or heterogeneous attributes of neighbor instances. Consequently, the resultant buffer zone exposes low accuracy and useless zone (information) and loses the useful zone (information).

To overcome these shortcomings of the existing buffer algorithm, an innovate algorithm called *generalized buffer algorithm* (*GBA*) is proposed. The basic idea of the GBA is to utilize homogeneous rules to induce the generation of a buffer zone for which buffer radius is variable upon the attributes of instances. Consequently, the size of the buffer zone is not a regular shape. The GBA mathematical model can be expressed by

$$CL\_B_j = \{p : d(p, A_j) \leq r_{CL}\} \tag{8.3}$$

where r_{CL} is the generalized buffer radius, which is constrained by homogeneous rules; CL_B_j is the instance set, in which the distances of all instances are less than or equal to r_{CL}; $d(p, A_j)$ is the distance between p and A_j. The generalized buffer radius r_{CL} is the function of the attributes of the instances and can be calculated by the homogeneous rules (*CLR*), i.e.,

$$r_{CL} = d[(x, y), CLR] \tag{8.4}$$

where $d[(x, y), CLR]$ is a distance function of the coordinates (x, y) of a boundary point that is obtained using homogeneous rules.

The flowchart of the GBA is depicted in Figure 8.2. Let each instance $A_i = (x_1, x_2, \ldots, x_N)^T$ in data set $A=(A_1, A_2, \ldots, A_d)^T$ be a vector representing instance-ID, spatial

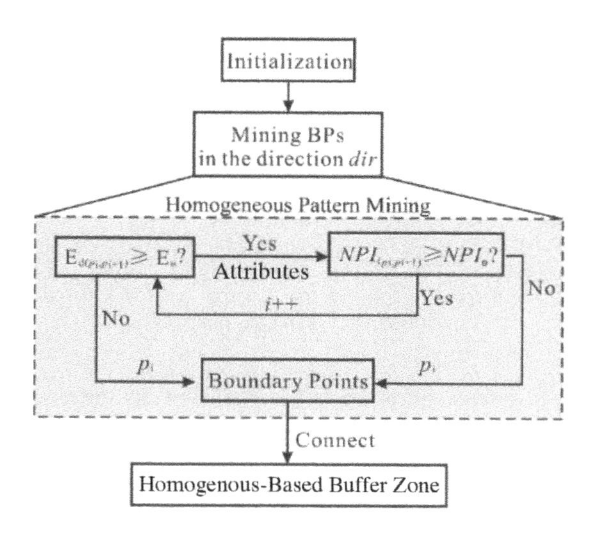

FIGURE 8.2 The flowchart of generalized buffering algorithm (GBA).

object type, location ∏, and attributes, etc., in which d is the number of object types, T is transpose, and location $\in$ spatial framework. The spatial object types contain point, line, and polygon entities. The basic steps for the generalized buffer generation are described as follow.

Step 1: Initialization: To build the generalized buffer zone of point, line, and polygon, the polar coordinate system is established, where the axis is horizonal and angle φ_i is the polar angle, varying from 0 to 2π. For point object, the pole is the itself. While, for line or polygon, the pole is the sampling points at the line or the arc of polygon.

Step 2: Determining buffer boundary points. This step consists of the following steps.

a. In the direction of φ, first, the Euclidean distance between instances p_i and p_{i+1} is calculated by

$$E_{d(p_i, p_{i+1})} = \sqrt{(x_i - x_{i+1})^2 + (y_i - y_{i+1})^2} \quad (8.5)$$

where $E_d(p_i, p_{i+1})$ is the Euclidean distance between instances p_i and p_{i+1}; x and y are the coordinates of instances. $E_d(p_i, p_{i+1})$ should be judged whether it is less than or equal to the threshold of distance E_θ by equation 8.6. If Yes, turn to (b). Otherwise, instance p_i is regarded as the buffer boundary point in direction of φ, and then turn to (c).

$$R(p_i, p_{i+1}) = \begin{cases} 1 & \textit{if } E_{d(p_i, p_{i+1})} \le E_\theta \\ 0 & \textit{if } E_{d(p_i, p_{i+1})} > E_\theta \end{cases} \quad (8.6)$$

where $R(p_i, p_{i+1})$ represents the R-proximity relationship between instances p_i and p_{i+1}. $R(p_i, p_{i+1})$=1 means that instances p_i and p_{i+1} are the candidate of homogeneous instances.

b. When instances p_i and p_{i+1} are determined to be the candidate homogenous instances, their attributes should be used to determine whether they are homogenous instances. A new participation index (*NPI*) is defined by

$$NPI(p_i, p_{i+1}) = \frac{similar\_nA(p_i, p_{i+1})}{total\_nA} \quad (8.7)$$

where $NPI(p_i, p_{i+1})$ is the participation index of instances p_i and p_{i+1}; $similar\_nA(p_i, p_{i+1})$ is the number of attributes in which instances p_i and p_{i+1} meet the same thresholds of attributes and $total\_nA$ represents the number of all attributes which attend the computation of *NPI*. For example, let $AT=\{AT_1, \ldots, AT_5\}$ be a set of attributes. For the attributes AT_1, AT_2, AT_4, and AT_5, the values of attributes of instances p_i and p_{i+1} meet the same thresholds of attributes, that is, $similar\_nA(p_i, p_{i+1}) = 4$, and $total\_nA = 5$. Consequently, *NPI* is 4/5. *NPI* is judged whether it is greater than or equal to the threshold of participation index NPI_θ. If Yes, turn to (a), and then the value of *i* automatically increases 1. If No, turn to (c), and instance p_i is regarded as the buffer boundary point in the direction of φ.

c. Collect buffer boundary points, and put them into a matrix in order. The direction line φ increase $\Delta\varphi$, which is the infinitesimal of φ. And then turn to (a), and initialize the value of *i*. If $\varphi = 2\pi$ and no homogenous instance is found, turn to (d).
d. Exit.

Step 3: Connecting buffer boundary points to create the generalized buffer zone: When all buffer boundary points are obtained, they are connected using THE interpolation method to form generalized buffer zone.

8.3 DISCUSSION OF THREE TYPES OF GBAS

The buffer generation usually consists of point, line, and polygon buffering algorithms. Thereby, three types of the generalized buffer generation algorithms for point, line, and polygon elements are described in the following sections, respectively.

8.3.1 Generalized Point Buffering Algorithms

The basic idea of the generalized point buffer generation is to determine the buffer radius of the neighbor instances using homogeneous rules with which the zone with the same attributes are clustered into the buffering zone (i.e., homogenous attributes). This implies that the buffer radius is no longer a constant, that is, it is a variable. The details are described as follows.

Step 1: Given a polar coordinate system where axis is horizonal and angle φ_i is the polar angle, varying from 0 to 2π, with the given definitions described in section 8.2 and the basic principle of the generalized buffer generation described in Chapter 8.3.1 above: for a target point object A_j (see Figure 8.3), A_j is taken as the point to be buffering and meanwhile the pole of the polar coordinates system, and $r\varphi$ represents the polar radius. The coordinates of boundary point ($BP(x, y)$) *of forming* the point buffer zone of target point object A_j can be defined by

$$P\_B_{A_j} = \left\{ p_i : E_d(p_i, A_j) \le r_\varphi \right\}, \quad \varphi \in [0, 2\pi] \tag{8.8}$$

$$r_\varphi = H\left\{ BP(x, y), CLR_\varphi \right\}, \quad \varphi \in [0, 2\pi] \tag{8.9}$$

$$CLR_\varphi = G\left\{ R(p_i, p_{i+1}), NPI(p_i, p_{i+1}) \right\}, \quad \varphi \in [0, 2\pi] \tag{8.10}$$

where P_B_{Aj} represents the point instance set that the distance between P_i and A_j is less or equal to buffer radius r_φ in the corresponding direction; CLR represents homogeneous rules; $H\{BP(x,y), CLR_\varphi\}$ is the function about the coordinates of boundary point ($BP(x, y)$) and CLR. $R(p_i, p_{i+1})$ represents that there exists R-proximity relationship between p_i and p_{i+1}. For a group of target point objects $A=\{A_1, A_2,\ldots, A_n\}$, the generalized point buffer is mathematically expressed by

$$P\_B_A = \bigcup_{i=1}^{n} P\_B_{A_i} \tag{8.11}$$

When the polar angle φ varies from 0 to 2π based on infinitesimal $\Delta\varphi$, in every corresponding orientation, homogeneous instances are mined until there are no new instances added into the homogeneous set and the buffer radius r_φ, which is the distance between the boundary point (the farthest homogeneous instance away from target point object A_j) and target point object A_j, will be determined.

According to the theory of a double integral, assuming that the boundary line is a continuous, differentiable, and integrable curve, the area of the point buffer zone can be expressed by

$$S\_B_{A_j} = \int_0^{2\pi} d\varphi \int_0^{r_\varphi} r dr \tag{8.12}$$

where S_B_{Aj} is the area of point buffer zone, $\Delta\varphi$ is the infinitesimal of polar angle φ, and r_φ is the buffer radius that changes with the change of boundary point's coordinates and subjects to homogeneous rules.

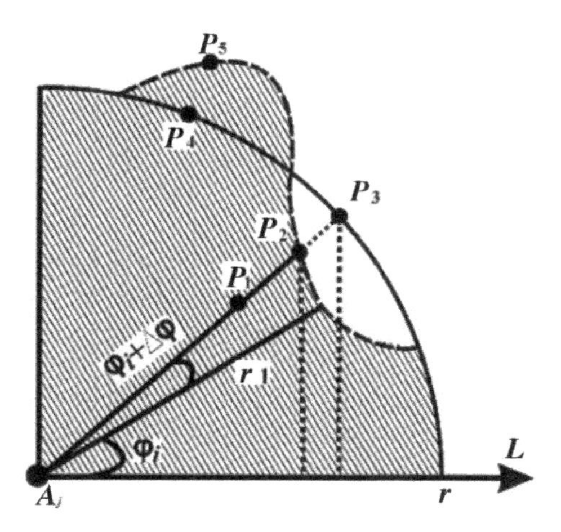

FIGURE 8.3 Generalized point buffer (GPIB) algorithm.

Step 2: Mining buffer boundary points using homogeneous rules, which consists of the following steps.

a. *Determination of homogeneous candidates.* In order to determine homogeneous candidates, the R-proximity relationship of instances must be determined first. According to equation 8.12, if the Euclidean distance between instances P_i and P_{i+1}, which are at the same direction line $\varphi_i + \Delta\varphi$, is less than or equal to distance threshold E_θ, there exists R-proximity relationship between these two instances. Therefore, those instances that satisfy R-proximity relationship constraint condition are the 2nd-order homogeneous candidates.
b. *Determination of prevalent homogeneous patterns.* When all 2nd-order homogeneous candidates are found, equation 8.13 is utilized to determine prevalent homogeneous patterns. According to equation 8.13, if the NPI of 2nd-order homogeneous candidate $\boldsymbol{C}_i$ is greater than or equal to the threshold NPI_θ, $\boldsymbol{C}_i$ is a 2nd-order prevalent homogeneous pattern. After this processing, only 2nd-order prevalent homogeneous are determined. Based on the generation method of k-th-order homogeneous candidate mentioned in the section 8.3, k-th-order prevalent homogeneous patterns will be obtained.
c. *Determination of buffer boundary points.* As shown in Figure 8.3, an example is made to explain the generation processes of the boundary point of the point-based buffer when the polar angle is $\varphi_i + \Delta\varphi$. According to the previous two steps, those instances (such as points P_1 and P_2) that satisfy homogeneous rules in this orientation are saved in a matrix, and the farthest homogeneous instance (for example point P_2) away from the target object instance A_j is regarded as the boundary point in this orientation and is put into the matrix of boundary point. Thus, the buffer distance $r\varphi_i$ in arbitrary polar angle φ_i can be mathematically expressed by

$$r_{\varphi_i} = \sqrt{X^2_{b\_point} + Y^2_{b\_point}} \tag{8.13}$$

When φ_i varies from 0 to 2π, all boundary points will be collected and put into a matrix in order.

Step 3: Generation of generalized point buffer zone. When all buffer boundary points are obtained, they can be connected using the interpolation method to form a generalized point buffer zone.

Compared to the implementation processes of a traditional buffer analysis algorithm, in the proposed algorithm, at the beginning, φ is increased in anticlockwise order based on infinitesimal $\Delta\varphi$. Meanwhile, instances are judged whether they are homogeneous in corresponding direction. And $n + 1$ boundary points can be produced, among which the first point and the last point are the same point, which makes the zone close. Finally, the cubic spline interpolation method is used to connect these boundary points, and the generalized point buffer zone is the area that is within the boundary. As shown in Figure 8.3, we can see that instance point P_3 which is the boundary point of the buffer zone in the traditional buffer zone, is replaced by point P_2 which is the new boundary point of the generalized point buffer zone, because instance point P_3 is not the homogeneous instance and instance point P_2 is the last homogeneous instance in this orientation.

8.3.2 Generalized Line Buffering

The basic idea of a generalized line buffer (GLB) algorithm is that a target line object is expressed by a point set $L_i = \{A_0, A_1, \cdots, A_m\}$ and is considered as an axis. The generation of the generalized line buffer (GLB) is mathematically expressed by

$$L_i\_B = \left\{ p : \bigcup_{j=0}^{m} [d(p, A_j) \le r_{A_j}] \right\} \tag{8.14}$$

$$r_{A_j} = D\left[BP_{A_j}(x_a, y_b), CLR_{A_j} \right] \tag{8.15}$$

where A_j is the j-th sample point along the target line object; r_{Aj} is the buffer distance corresponding to A_j, whose size varies upon the attributes of neighbor instances; Li_B is the instance set in which the distance between p and A_j is less or equal to r_{Aj} in the normal direction of the j-th sampled point along the target line object; $D[BP_{Aj}(x_b, y_b), CLR_{Aj}]$ is the function of the coordinates of boundary points, which is based upon the attributes of the neighbor instances decided by the homogeneous rule. CLR_{A_j} represents the induced homogeneous rules (CLR) in the corresponding normal direction of the target line object, which regards the j-th sample point as the foot point. For multiple target line objects $L=\{L_0, L_1,\ldots, L_n\}$, the generalized line buffer can be defined as

$$L\_B = \bigcup_{i=1}^{n} L_i\_B \tag{8.16}$$

In order to calculate the area of the generalized line buffer zone across two endpoints of line, vertical lines of line are made, such as $V_1(x)$ and $V_2(x)$, respectively. Let d_s and d_r represent the infinitesimal element of line object and buffer radius, respectively. And let S_{A0} and S_{Am} be the areas of zone that are formed by the buffer zone of endpoints and vertical lines, respectively. The area of the generalized line buffer zone can be calculated by

$$LS = \int_0^{L_i} ds \int_0^{r_{A_j}} dr + S_{A0} + S_{Am} \tag{8.17}$$

where r_{Aj} can be got from equation 8.15, which is constrained by homogeneous rules; S_{A0} and S_{Am} can be acquired based on equation 8.12.

When all boundary points are found, they will be connected by using the cubic spline interpolation method to form an enclosed zone (see Figure 8.4). Compared to the traditional line buffer algorithm, the proposed method considers that the buffer distance is no longer a constant but is a variable that is decided by the attributes of neighbor instances and decided by homogeneous rules. As a result, the produced line buffering zone is much closer to the actual situation.

With the algorithm proposed here, the details of the generation of GLB for a target line object can be summarized as the following steps.

Step 1: Initialization of line: First, the target line object is virtually sampled and noted as a series of points. Let *ss* be a sampling distance; a series of sampling points are obtained along the target line and is expressed by $L_i = \{A_0, A_1, \ldots, A_7\}$ (see Figure 8.4). As shown in Figure 8.4, A_0, A_4, and A_7 are the starting point, inflection point, and endpoint, respectively.

Step 2: Determination of the rectilinearity of adjacent three points. Assume that coordinates of the adjacent three points are $A_{i-1}(X_{i-1}, Y_{i-1})$, $A_i(X_i, Y_i)$, and $A_{i+1}(X_{i+1}, Y_{i+1})$, their vectors are expressed by $\overrightarrow{A_{i-1}A_i}$, $\overrightarrow{A_iA_{i+1}}$. The vector product of the two vector is used to determine their rectilinearity, that is,

$$\overrightarrow{A_{i-1}A_i} = (X_i - X_{i-1}, Y_i - Y_{i-1}) = \vec{g} = (m_1, n_1) \tag{8.17}$$

$$\overrightarrow{A_iA_{i+1}} = (X_{i+1} - X_i, Y_{i+1} - Y_i) = \vec{f} = (m_2, n_2) \tag{8.18}$$

The normal vector is calculated by

$$\vec{N} = \overrightarrow{A_{i-1}A_i} \times \overrightarrow{A_iA_{i+1}} = \vec{g} \times \vec{f} = (m_1n_2 - m_2n_1)\vec{j} = \beta\vec{j} \tag{8.19}$$

where $m_1=(X_i-X_{i-1})$, $n_1=(Y_i-Y_{i-1})$, $m_2=(X_{i+1}-X_i)$, $n_2=(Y_{i+1}-Y_i)$; $\vec{j}$ is the unit vector that is perpendicular to the plane constructed by vectors $\overrightarrow{A_{i-1}A_i}$ and $\overrightarrow{A_iA_{i+1}}$. If β is

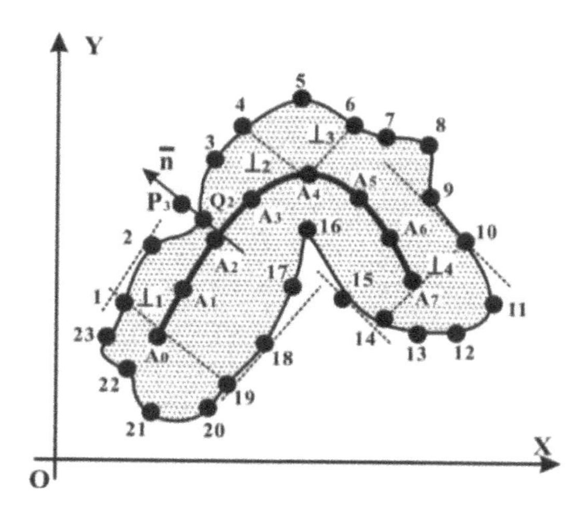

FIGURE 8.4 Generation of generalized line buffer zone.

equal to zero, then $\overrightarrow{A_{i-1}A_i}$ and $\overrightarrow{A_iA_{i+1}}$ are collinear; for example, points A_1, A_2, and A_3 in Figure 8.4 almost lay a straight line; the β value is close to zero; if β is greater than or less than zero, for instance, points A_3, A_4, and A_5, the middle point A_4 is usually an inflection point. Furthermore, the concavity and convexity of inflection point should be judged. When β determined by the products of the two vector is greater than zero, if the former vector turns to the latter vector with the minimum angle at a counterclockwise direction, the inflection point is a convexity point; in contrast, the inflection point is a concave point.

Step 3: Generation of line buffering: After linearity, inflection, and concavity–convexity of inflection point at the sampled points along a target line are completed, the following work is the generation of GLB using the homogeneous rule. First spatial and nonspatial attributes are selected to determine a rough candidate for the zone using the homogeneous rule. The mathematical model is expressed by

$$NS\_A_{i,j} = \begin{cases} NS\_A_{i,j} & NS\_A_{i,j} \text{ satisfies threshold} \\ NAN & NS\_A_{i,j} \text{ does not satisfy threshold} \end{cases} \quad (8.20)$$

where $NS\_A_{i,j}$ is the value of a certain nonspatial attribute and i and j are coordinates. With implementing this step, those instances without meeting the given threshold will be excluded. Afterward, the generation of GLB is carried out, for which three types of sampled points are considered as follows, respectively.

a. *When the points are neither at inflection points nor at endpoints.* Taking point A_2 in Figure 8.5 as an example to explain the line buffering generation, assume that $\vec{n}$ is a normal lines at line L_A, the instances points A_2, Q_2, and P_3 lie on the normal line $\vec{n}$. The Euclidean distance between them is

$$D_{A_2,Q_2} = \sqrt{\left(X_2 - X_{Q_2}\right)^2 + \left(Y_2 - Y_{Q_2}\right)^2} \quad (8.21)$$

If $D_{A2,Q2}$ is less than or equal to a given threshold D_θ, A_2 and Q_2 exist in an R-proximity relationship. Thereby, they are considered as candidate of a homogeneous set. When all instances of candidate homogeneous pattern $\{A, Q\}$ are determined, according to equation 8.20 and equation 8.21, the participation index (PI) is applied to decide whether the candidate homogeneous pattern is prevalent homogeneous pattern or not. If NOT, it is not considered as homogenous. When k-th order homogeneous patterns are mined, a k + 1-order homogeneous pattern may be found. This homogeneous mining process will not be stopped until no qualified homogeneous pattern is found. The point belongs to boundary point, that is point Q_2 in Figure 8.4.

b. *When the points are at inflection*. Taking point A_4 in Figure 8.4 as an example to explain the line buffering generation, two perpendicular lines are made, noted as $\perp_2$ and $\perp_3$, along the adjacent two line segments (see line segments A_3A_4 and A_4A_5), and then the parallel line of directed line segments (A_3A_4 and A_4A_5) crossing the buffer zone's boundary point (for example points 3) corresponding to the adjacent points (i.e., point A_3) of point A_4 are made. Then at the convex side, an arc is constructed, which makes point A_4 as the center and regards the distance between point A_4 and the parallel line as the radius. The arc intersects with $\perp_2$ and $\perp_3$ to form a closed zone, which is the initialization zone for mining the homogeneous pattern. For inflection point A_4, R-proximity neighbor relations between instances will be determined by equation 8.12 in the closed zone, and homogeneous patterns are mined in this initialization zone. Finally, the buffer zone's boundary points, such as points 4, 5, 6, will be acquired. However, at the side of concave, R-proximity neighbor relations and homogeneous patterns are mined on the angular bisector of $\angle A_3A_4A_5$. Then the boundary points corresponding to inflection points are determined.

When the points are at endpoints. Taking point A_7 in Figure 8.4 as an example to explain the generation of GLB, crossing the endpoint, a vertical line of a broken line is made, and parallel lines of broken line are produced across the buffer zone's boundary points (such as points 9 and 15) corresponding to point A_6, which is adjacent to endpoint A_7, and then an arc is made, which makes the midpoint of the vertical line segment $\perp_4$, which intersects with the two parallel lines as the center and the distance between the midpoint and parallel line as the radius. The arc intersects with the two parallel lines, respectively. The vertical line segment $\perp_4$ and the arc form a closed zone which is the initialization zone for mining homogeneous patterns. For the endpoint, the process of mining homogeneous patterns is the same as step (a). Finally, the boundary points of a generalized line buffer zone (such as points 9, 10, 11, 12, 13, and 14) are collected and put into the set.

8.3.3 Generalized Polygon Buffering

Polygon-based buffer algorithm includes outwards or inwards buffer radius to generate a buffer zone. If the boundary of a polygon is considered as an enclosed curve, the polygon-based buffer algorithm is somewhat similar to the line-based buffer algorithm (see Figure 8.5). The mathematical models can be referenced to equation 8.20 and equation 8.21. The implementations can be carried out by:

Step 1: Extraction of the polygon boundary, and initialization of the boundary line. Because the buffer of the polygon makes the boundary line the axis and a given distance as the buffer distance to dilate outside or shrink inside, so the initialization process is similar to the broken line's. Points that are on the boundary line should be collected by applying step size. Although the first point and the last point overlap (such as points Q_0 and Q_{16}

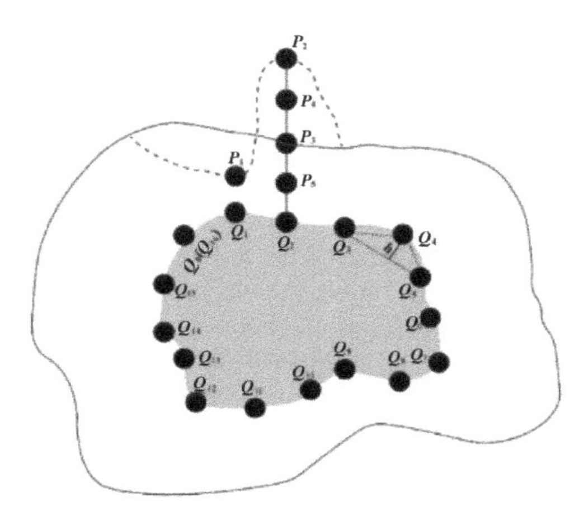

FIGURE 8.5 Generalized polygon buffer (GPLB).

in Figure 8.5), it is not necessary to separately consider the buffer boundary of terminal points, as long as it is according to the generalized buffer construction method of the middle point of three consecutive points.

Step 2: Determination of the rectilinearity of adjacent three points. The polygon boundary is a curve, but in this chapter, directed line segments are connected to replace the curve (such as $\vec{a}$, and $\vec{b}$ in Figure 8.10). When the amount of line segments is more enough, the boundary of the closed zone, which is constructed by directed line segments, is smoother. The rectilinearity of three adjacent points (such as points Q_3, Q_4, and Q_5) can be determined by

$$h = \frac{(Dx_4 + Jy_4 + K)}{\sqrt{D^2 + J^2}} \tag{8.22}$$

Where h is the distance between middle point Q_4 and the straight line across the first point Q_3 and last point Q_5 of three adjacent points; D, J, and K are coefficients of equation of the straight line; x_i and y_i $(i = 4)$ are coordinates of the middle point (for example point Q_4). If h_i is less than or equal to the threshold, then the adjacent three points can be regarded as collineation.

Step 3: Creation of generalized polygon buffer. Similar to the operation for line in section 8.3.3, the points on the boundary line of polygon need to be judged whether they are inflection points. If h_i is greater than the threshold, then the ***i***-th point is inflection point. When the types of points are determined, homogeneous rule should be utilized to guide the generalized polygon buffer zone. First, nonspatial attributes are selected to determine rough candidate homogeneous instances using equation 8.14, and then in order to get candidate homogeneous patterns, *R*-proximity relations among these rough candidate homogeneous instances should be determined in a normal line direction by Euclidean distance, and the points that are on the boundary of the polygon but not the inflection points are made as the foot of perpendicular. For instance, in Figure 8.11b, P_j is a point on the boundary line of the polygon, and there are four points that satisfy the Euclidean distance condition, that is, there are *R*-proximity relationship among them in a normal line direction that make Q_j the foot of perpendicular. Then 2-order candidate homogeneous patterns can be acquired, such as $\{Q_j, P_i\}$ and $\{P_i, P_{i+1}\}$. For the sake of determining whether these candidate homogeneous patterns are prevalent homogeneous patterns, participation index ($NPI\{C\}$) of patterns should be calculated by equation 8.1 and equation 8.2. If $NPI\{C\}$ is greater than or equal to the threshold, then pattern C is the prevalent homogeneous pattern. In addition, k-th order homogeneous patterns can be generated based on $k-1$-th order and $k-2$-th order homogeneous patterns. At last, the point (for example point P_2) that meets homogeneous rules and is farthest away from Q_j in a normal line direction is regarded as the boundary point of the generalized polygon buffer zone. The same operation is performed for other points that are on the boundary line of the polygon. All boundary points of the generalized polygon buffer zone are collected and put into a set in order.

8.4 EXPERIMENTS AND ANALYSIS

Traditional buffer analysis usually took the noise pollutions as the paradigm. This chapter also takes the noise pollutions paradigm in two cities of China as an example to explain how the generalized buffer is used for spatial analysis and what difference is between the generalized and traditional buffer algorithm.

8.4.1 Data Sets

Data set-1: The investigation area is the electronic vector data of Bao'an District, Shenzhen, China, obtained from the Geographical Information Monitoring Cloud Platform in 2015 with an ArcGIS shp file. Data set-1 contains more than 30 million pieces of information (e.g., hospitals, buildings, supermarkets, gas stations, banks, etc.), in which there are 24 types of point features consisting of 1,876 point elements, (e.g., airports, parking, gas stations, etc.), 11 types of line features consisting of 26,135 line elements (e.g., in-state free highway, provincial highway, pedestrian paths, etc.), and seven types of polygon features consisting of 1,201 polygon elements (e.g., provincial boundaries, regional boundaries, lakes, etc.).

Data set-2: The investigation area is the electronic data of Beijing in 2015 and is obtained from the Geographical Information Monitoring Cloud Platform, with ArcGIS shp format. Data set-2 contains more than 30 million pieces of information (e.g., hospitals, buildings, supermarkets, gas stations, banks, etc.). Data set-2 contains 18 types of point features consisting of 124,252 point elements (e.g., village, school, bus station, etc.), nine types of line features consisting of 112,057 line elements (e.g., expressways, provincial highways, pedestrian paths, etc.), and four types of polygon features consisting of 6,933 polygon elements (e.g., lakes, rivers, boundaries of regions, etc.).

In the two data sets, the point noise source is assumed from in-situ construction, such as excavating, blasting, cement mixing, etc. The line noise source is assumed from moving vehicles on the highway. The polygon noise source is assumed from manufacturing, in which the machines are operated continuously. Many buildings are surrounding these noise sources, i.e., surrounding the point element, line element, and the polygon element.

- *Spatial data:* The spatial data for describing point, line, and polygon features are X, Y coordinates. The datum and the projection coordinate system for spatial data for Data set-1 and Data set-2 are listed in Table 8.1.
- *Attribute data:* The attribute data of point, line, and area features include quantitative and qualitative attributes (see Table 8.2). In order to investigate the extent of noise propagation using point, line, and polygon buffer algorithms without losing generality, only one attribute, namely, integrated areal density (IAD), is selected, and other attributes are not taken into account in this step. This assumption is consistent with Kang (2004), who though that the IAD of a building hinders noise propagation.

TABLE 8.1
The Information of Spatial Data

Attributes	Data Set-1	Data Set-2
X, Y coordinates	Geographical coordinate (system: Lat Long for MAPINFO type) Datum: D_MAPINFO Prime meridian: Greenwich Angular Unit: Degree Spheroid: World_Geodetic_System_of_1984 Semimajor Axis: 6378137.00 Semiminor Axis: 6356752.314 Inverse Flattening: 298.257	Geographical coordinate (system: GCS_WGS_1984) Datum: D_WGS_1984 Angular Unit: Degree (0.017453292519943299) Prime Meridian: Greenwich Spheroid: WGS_1984 Semimajor Axis: 6378137.00 Semiminor Axis: 6356752.314 Inverse Flattening: 298.257

TABLE 8.2
Instance Nonspatial Attributes

Type of data		Nonspatial attribute
Point, Line, Polygon	Quantitative attributes	Time, Length, Width, Height, Area, Quantity, Size, Level, Density, Integrated areal density (IAD)
	Qualitative attributes	Name, Category, Number, Utilization type, Characteristics

8.4.2 Generalized Buffer Analysis

The experiments were conducted on a computer with an Intel Xeon E5645 Six core 2400 MHz (12MB Cache) and 4GB of RAM.

8.4.2.1 Experiment for Data Set-1

8.4.2.1.1 Generalized Point Buffer Analysis

The purpose of studying noise pollution using point buffer analysis is to investigate how big an area is impacted by a noise source. Usually, the volume of a sound higher more than 80 dB is considered as noise and of lower than 45 dB–60 dB is NOT considered as urban environmental noise. The noise attenuation is expressed

$$\Delta dB = 10\log\left(\frac{1}{4}\pi r^2\right) \tag{8.23}$$

where r is the noise propagation distance (m) and ΔdB is the noise attenuation (dB), which decreases with increasing distance square as a logarithmic function.

This chapter selects 45 dB as a threshold of either sound or noise, that is, bigger than 45 dB as noise. With the threshold, the buffer radius in traditional point buffering algorithm is

$$r = \sqrt{4\times 10^{(80-45)/10}/3.1415926} = 63.5(\mathrm{m}) \tag{8.24}$$

In order to compare the traditional point buffering algorithm (TPIA), we take point A, which is a mechanical manufacturing factory, as the noise source to create a point buffer analysis, as depicted in Figure 8.6. As observed from Figure 8.6, the traditional point buffer area is a circle centered at Point A, i.e., the noise source. This means that traditional point buffer analysis thinks that the noise propagation is along a flat terrestrial area in all directions without any obstacle. In fact, the noise propagation is impacted by many factors, such as buildings, and as a result, the noise propagation is NOT a complete circle; that is, the noise propagation distances are NOT equal in all directions.

In the generalized point buffer (GPIB) algorithm, also taking Point A as the noise source, 45 dB as a threshold of either sound or noise, buildings noted as Point P_i,

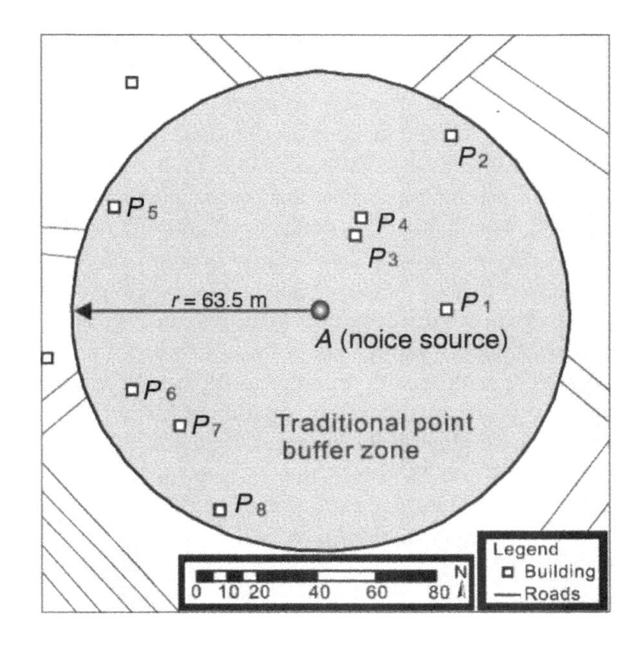

FIGURE 8.6 Traditional point buffer zone with a buffer radius of 63.5 m.

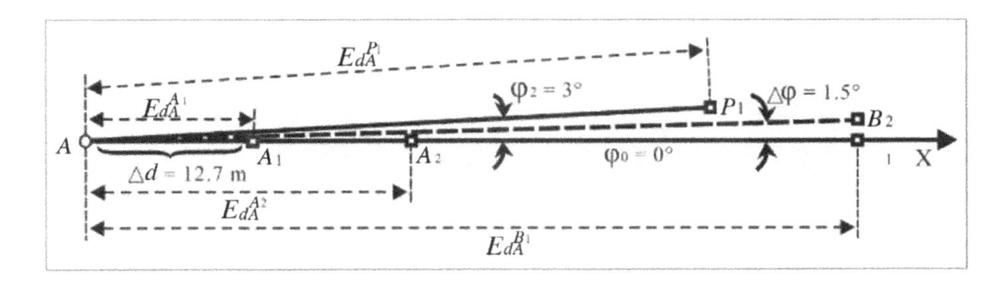

FIGURE 8.7 Generalized point buffer analysis algorithm.

($P_i = P_1, \ldots, P_8$) in Figure 8.6 as obstacles, whose attributes are listed in Table 8.3, IAD as factor of noise attenuations. The steps of the GPIB algorithm are as follows.

Step 1: Establish a polar coordinate system. Point A is taken as the pole, and easting as polar axis, noted, X-axis. φ is defined as the "*polar angle*", whose value is determined relative to the X-axis counterclockwise, $\Delta\varphi$ is defined as the angle increment at a counterclockwise rotation. In this experiment, the polar angle is expressed using φ_i, that is, $\varphi_{i+1} = \varphi_i + \Delta\varphi$, and Δd is defined as increment distance (distance step) along rotational axis starting from pole Point A (see Figure 8.7).

Step 2: Calculating the sound propagation course. Starting at Point *A*, along the X axis, that is, $\varphi_0 = 0°$, set a distance increment of 12.7 m (the 63.5/5 = 12.7 m, noted as Δd=12.7 m; other lengths are discussed later). The starting point is located at Point A_1. The Euclidean distance between Point *A* and Point A_1 can be calculated using equation 8.5, noted as $Ed(A, A_1)$ = 12.7, and noise attenuation is also calculated using equation 8.23, which is 20.8 dB. Meanwhile, it is determined whether the sound propagation encounters any building. It has been indicated from the data set that no buildings have blocked the nose propagation (see Figure 8.7). Because $Ed(A, A_1)$ is less than the distance threshold, noted as $E_\theta = 63.5$ m, Point A_1 is considered to be homogenous with Point *A*, that is, Point A_1 with Point *A* has an *R-proximity relationship.* In order to further determine their homogeneity, the *NPI* is calculated using equation 8.7, where the *total_nA* in equation 8.7 is the number of all attributes, which is 3 (IAD and XY coordinates) in Table 8.1 and Table 8.2, that is, *total_nA* = 3. The *similar_nA* is the number of attributes for each point (see Table 8.2). Point *A* has 2 attributes (XY coordinates), *similar_nA*=2. that is, NPI_A=2/3. Therefore, the threshold of *NPI* is calculated using equation 8.7, that is, $NPI_\theta = 2/3$. When the *NPI* value of the candidate instance is equal to or greater than NPI_θ, this means that the candidate instance is homogeneous with Point *A*. On the other hand, an *NPI* value of Point A_1 is $NPI_{A1} = 2/3$, which is equal to NPI_θ. This means that Point A_1 and Point *A* are homogeneous. Moreover, the noise value at Point A_1 is 59.2 dB (≥ 45 dB), so the noise at Point A_1 will continue to propagate along the X-axis.

Step 3: Repeating the same operation at the same distance increment, that is, (Δd = 12.7 m), the sound is propagated to Point A_2; calculate the Euclidean distance between Point *A* and Point A_2 and noise values, and also determine whether encountering any building. With calculations, the volume of sound is 27.0 dB, and no building is encountered. The Euclidean distance of Point A_2 and Point *A* is $Ed(A, A_2)$ = 12.7 m + 12.7 m = 25.4 m, which is less than $E_\theta = 63.5$ m and $NPI_{A2} = 2/3$, which is equal to NPI_θ. This means that Point *A* and Point A_2 are homogeneous. In addition, the noise value at Point A_2 is 53 dB (≥ 45 dB). This means that the sound will continuously propagate along the X-axis. Repeat *Step 2* until the volume of sound is equal to and less than 45 dB, and/or encounters the buildings which can block sound propagation. Therefore, a boundary point, far away (63.5 m) from Point *A* is found and noted, Point B_1, and $Ed(A, B_1)$ = 63.5 m, which is the boundary of the GPIA at $\varphi_0 = 0°$ (see Figure 8.7).

Step 4: Anticlockwise rotating the polar axis with an increment angle of 1.5°, that is, $\Delta\varphi = 1.5°$ and $\varphi_1 = \varphi_0 + \Delta\varphi = (0° + 1.5°) = 1.5°$ (see Figure 8.7). Repeating *Step 2* and *Step 3* for the same operations along new X-axis at polar angle of 1.5°. With the repeated operations and the same calculations, during which there are no buildings/blocks, until the boundary point B_2 is recognized.

Step 5: Repeat *Step 4* with anticlockwise rotating of the polar axis at an increment angle of 1.5°, that is, $\Delta\varphi = 1.5°$, $\varphi_2 = \varphi_1 + \Delta\varphi = (1.5° + 1.5°) = 3°$ (see

Figure 8.7). Repeating *Step 2* and *Step 3* along the X-axis at a polar angle of 3°, it is found that the sound propagation meets a blocking building, noted as Point P_1. Calculate the Euclidean distance from Point A to Point P_1 using equation 8.5, that is, $Ed(A, P_1) = 44$ m, which is less than $E_\theta = 63.5$ m. This implies that Point P_1 is homogenous with Point A, since it satisfies the *R-proximity relationship* constraint condition in equation 8.6. In addition, for Point P_1, NPI_A = (similar_nA)/(total_nA) = 1/3, which is smaller than the threshold, $NPI_\theta = 2/3$. Therefore, it can be concluded that Point P_1 and Point A are not homogeneous. This implies that the sound propagation ends at Point P_1. Thereby, Point P_1 is considered as the boundary point of the GPIB algorithm.

Step 6: Repeating *Step 2* through *Step 5* with anticlockwise rotating the polar axis with an increment angle of 1.5°, until $\varphi = 360°$, the GPIB algorithm is finished. The result is shown in Figure 8.8.

Comparison analysis with the different angle increments: In order to compare the accuracy of the GPB algorithm, the different increments of angle are set up at 3°, 6°, and 9°. The same operations are carried out as ones previously described, and the results are shown in Figures 8.9a, 8.9b, and 8.9c, respectively.

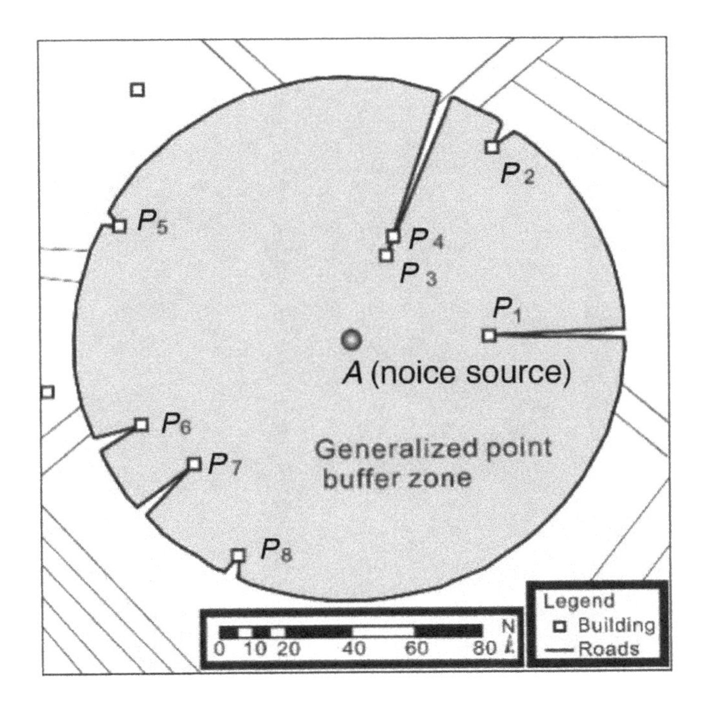

FIGURE 8.8 Generalized point buffer using Data set-1 with the increment angle at $\Delta\varphi = 1.5°$.

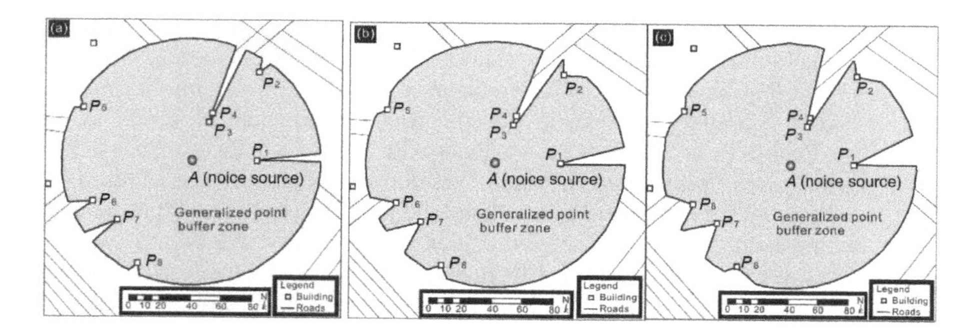

FIGURE 8.9 Generalized point buffer using Data set-1 with the different increment angles at (Fig. 8.15a) $\Delta\varphi_2 = 3°$, (Fig. 8.15b) $\Delta\varphi_3 = 6°$, and (Fig. 8.15c) $\Delta\varphi_4 = 9°$.

8.4.2.1.2 Generalized Line Buffer Algorithms

The validation takes a vehicle's noise along the highway as a noise source to investigate how a big area is impacted by the vehicle noise source using the generalized and traditional line buffer algorithm. In order to explain the GLB algorithm in detail, two endpoints and one inflection point are selected as examples. Also, assuming that the vehicles' noise is 80 dB (vehicle noise is usually between 60–80 dB (Ma and Chang 2018)), the traditional line buffer zone is generated using a buffer distance of 63.5 m along the highway (line element). The result is depicted in Figure 8.10.

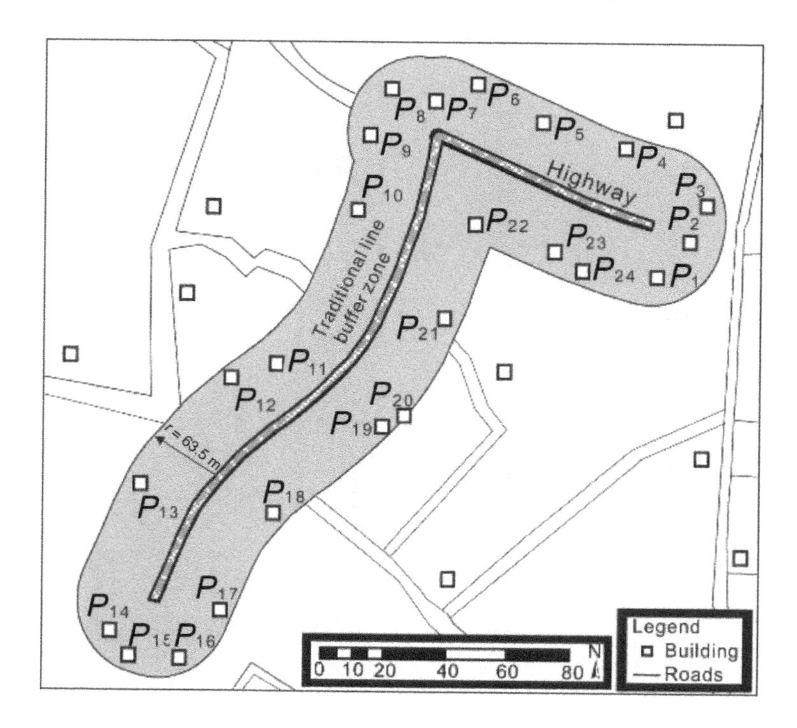

FIGURE 8.10 Traditional line buffer analysis with a buffer distance of 63.5 m.

Similar to the GPIB algorithm, the GLB algorithm also takes the buildings – which block the noise propagation upon their properties (e.g., IAD) – as obstacles, and they are represented by Point P_i (i = 1, . . ., 24) in Figure 8.10. The steps for the GLB algorithm are described as follows.

Step 1: The line is first resampled into the line segments at a given distance increment, noted $\Delta L = 1m$, without losing generality. The points are called *sample points.* The six sample points, noted as A_i (i = 1, ..., 6), located in the starting point and the inflection of the line (highway), respectively, are taken as examples to explain the process of the GLB algorithm (see Figure 8.11a). The spatial attributes and nonspatial attributes are listed in Table 8.2 and Table 8.3.

Step 2: Start from Point A_1 (see Figure 8.11b), determining whether Point A_1 is either endpoint, or inflection point, or neither using the method in Chapter 8.4.2.1.2. Because Point A_1 is an endpoint, a line perpendicular to the line passing Point A_1 is made, that is, $B_1A_1B_2$, where Point B_1 and Point B_2 are the boundary points of the noise propagation in the GLB algorithm. Similarly, the same calculation and process are made for Point A_1 using the method described in *Step 3* of the GPIB algorithm in Section III.B. With the calculated results, it is found that the Point B_0 and Point B_2 are traditional boundary points without encountering any buildings (blocks). Thereby, their Euclidean distances, $Ed(A_1, B_0)$ and $Ed(A_1, B_2)$, are 63.5 m, respectively.

Step 3: With an increment angle of $\Delta\varphi$ = 1.5°, and rotating anticlockwise the axis AB_1^0 until φ = 1.5°, set an increment distance of 12.7 m along the axis $A_1\ B_1^0$ at B_1^0 and repeat *Step 2* through *Step 5* in the GPB algorithm described in Section III.B to determine whether any building (block) is encountered during the noise propagation. When the noise propagates to Point b, it is found that no buildings (blocks) are encountered with Point b. With the repeated operation, no buildings (blocks) are encountered until reaching Point B_1^1; thus, Point B_1^1 is considered as a boundary point.

Step 4: Repeat *Step 3* with an increment angle of $\Delta\varphi$ = 1.5° until the rotation angle reaches φ = 31.5°; it is found that the sound propagation encountered a building, noted as Point P_1. Calculating the Euclidean distance from Point A_1. to Point P_1 using equation 8.5, that is, $Ed(A_1, P_1)$, it satisfies the *R-proximity relationship* constraint in equation 8.6 (that is, less than 63.5 m) and is considered to be homogenous with and a candidate for being homogeneous with Point A_1. Further calculate NPI_{P1} = $(similar\_nA)/(total\_nA)$ = 1/3, which is smaller than the threshold NPI_θ = 2/3, to determine whether Point P_1 is homogeneous with Point A_1 or not using equation 8.7 in Section III.B; with the calculated result, it is found that Point P_1 and Point A_1 are NOT homogenous. Thus, Point P_1 is considered as the boundary point of the GLB zone.

Step 5: Continue increasing the increment angle of $\Delta\varphi$ = 1.5° and repeat *Step 4*; it is found that Point P_2 and Point P_3 are NOT homogeneous with Point A_1, which means that they are boundary points.

Step 6: Continue increasing the increment angle of $\Delta\varphi$ = 1.5°, repeat these steps until Point B_2.

Step 7: With the assumed length increment, $\Delta L = 1.0$ m, Point A_2 is operated to determine whether it is either an endpoint, an inflection point, or neither by the methods described in Section II.C. With the computation, Point A_2 is none of them. Two line-segments (A_2B_3, A_2B_4) are made, respectively, on both sides of the highway (line element); meanwhile, the homogeneous points along both of the line segments, A_2B_3, A_2B_4, are determined (see Figure 8.11b). It can be concluded that no buildings are encountered in both of the line segments, that is, Point B_3 and Point B_4, as the boundary points.

Step 8: Repeat the same operation with the same length increment, until Point A_3, it is found that the sound propagation encounters a building, noted, Point P_4. Calculating the Euclidean distance between Point A_3 and Point P_4 using equation 8.5, that is, $Ed(A_3, P_4) = 37$ m, which satisfies *R-proximity relationship* constraint condition in equation 8.6 (that is, less than 63.5 m), and are considered as a candidate of homogeneous with Point A_3. Further

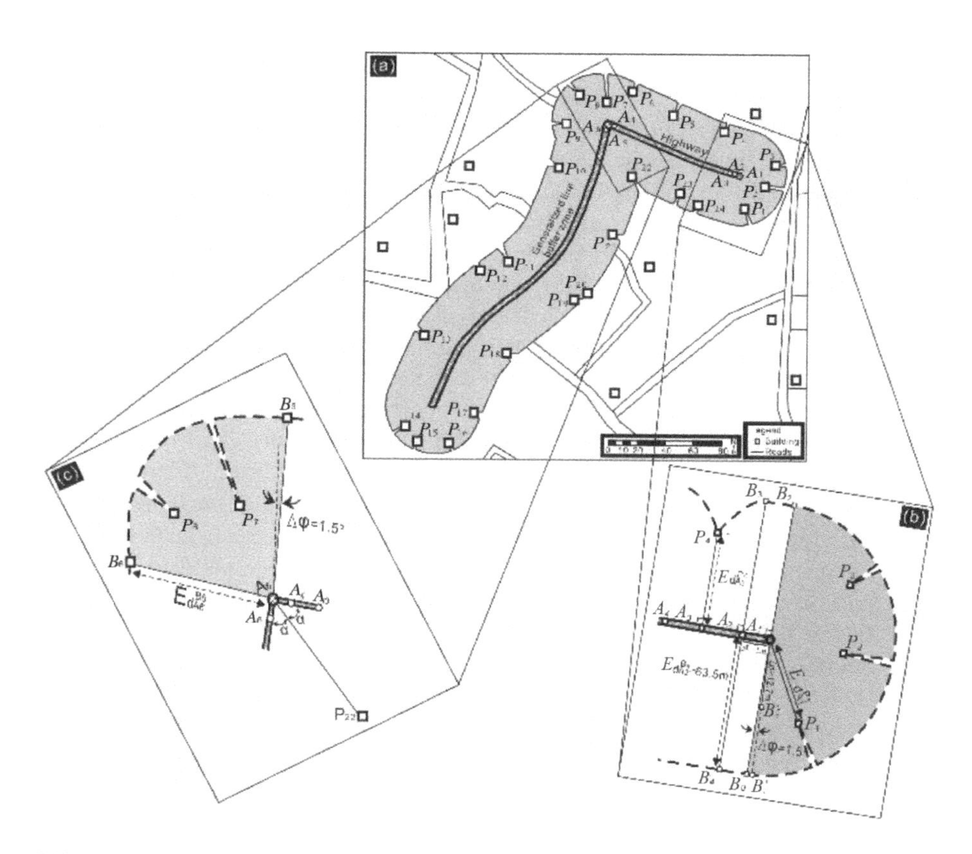

FIGURE 8.11 (a) Generalized line buffer using Data set-1 with the increment length at ΔL 1.0 m; (b) and (c) generalized line buffer analysis algorithm implementation step diagram.

determining whether Point P_4 is homogeneous with Point A_3 using *Step 4*, Point P_4 is considered as the boundary point of the generalized buffer analysis. Similarly, Point B_5 is a boundary point.

Step 9: Repeat *Step 7* and *Step 8* up to Point A_5 (see Figure 8.11c). With the computation, Point A_5 is an inflection point by the methods described in Section II.C. Two vertical lines along two line segments (A_4A_5, A_5A_6) are made. The boundary points are determined using the same *Step 3*. Point B_6 and Point B_7 are boundary points. Their Euclidean distances, $Ed(A_5, B_6)$ and $Ed(A_5, B_7)$, are both 63.5 m.

Step 10: Assuming that Point B_6 is a starting point, repeat *Step 4* with an increment angle of $\Delta\varphi = 1.5°$ to determine the boundary points. Point P_7 and Point P_8 are selected.

Step 11: Continue increasing the increment angle of $\Delta\varphi = 1.5°$; repeat the steps given until Point B_7.

Step 12: On the angular bisector of $\angle A_4A_5A_6$, repeat the same *Step 4*, with Point P_{22} as the boundary point.

Step 13: Connect all boundary points to form a generalized line buffer analysis area. The results are shown in Figure 8.11a.

Generalized line buffer generation using different length increments ΔL: In order to compare the difference when using different distance increments, ΔL, the difference value of ΔL at 5 m, 10 m, and 15 m are set. Repeating the same steps described, the results of the GLB zones with the different length increment are depicted in Figure 8.12a, Figure 8.12b, and Figure 8.12c.

8.4.2.1.3 Generalized Polygon Buffer (GPLB) Generation

Assume that the noise source is produced by a factory at 80 dB (Xia et al. 2001). The region of the factory with noise is depicted in Figure 8.13. Without losing the generality, the noise at any point within the factory is the same. The traditional polygon buffer zone is generated using a buffer distance of 63.5 m and is shown in Figure 8.13.

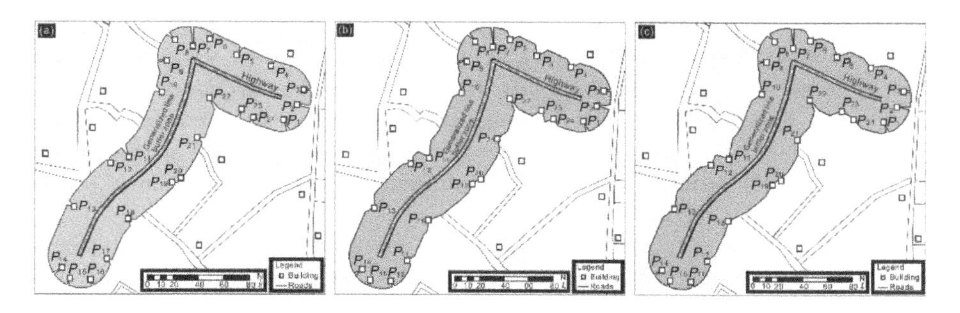

FIGURE 8.12 Generalized line buffer using Data set-1 with the different increment length at (a) $\Delta L_2 = 5$ m, (b) $\Delta L_3 = 10$ m, and (c) $\Delta L_4 = 15$ m.

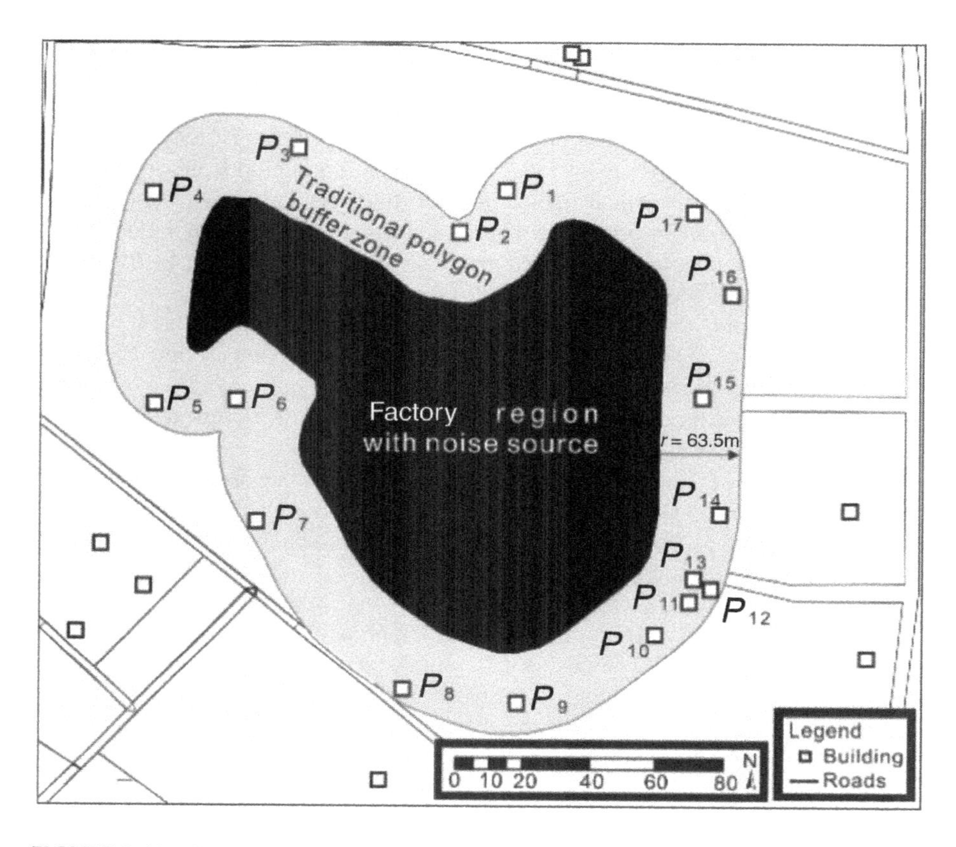

FIGURE 8.13 Traditional polygon buffer analysis with a distance of 63.5 m.

The operations of the GPLB algorithm are almost the same as the one of the GLB; the unique difference is the starting point is the same as the endpoint. For this reason, any point at a curve is selected as the starting point. For example, Point Q_1 is selected as the starting point, and the boundary of the polygon is resampled into curve segments starting from Point Q_1 at a given curve of 1.0 m, noted as $\Delta S = 1.0$ m. The points are called *sampled points*. The four sampled points, noted as Point Qi ($i = 1, \ldots, 4$), which are located on the boundary of a polygon (factory with noise) are taken as examples to explain the algorithm of the GPLB zone generation (see Figure 8.14). The spatial attributes and nonspatial attribute for each of the sampled points are the same as the ones described earlier.

For Point Q_2 (see Figure 8.14), whether is it an inflection point or not is determined by the method described in Chapter 8.4.1, with which Point Q_2 in an inflection point. The results are shown in Figure 8.14.

GPB Zone Generation using different curve increments ΔS: In order to investigate the accuracy of the GPB algorithm with the different increment of curves, ΔS, the increment of curves at 5 m, 10 m, and 15 m are set up for the experiments. The

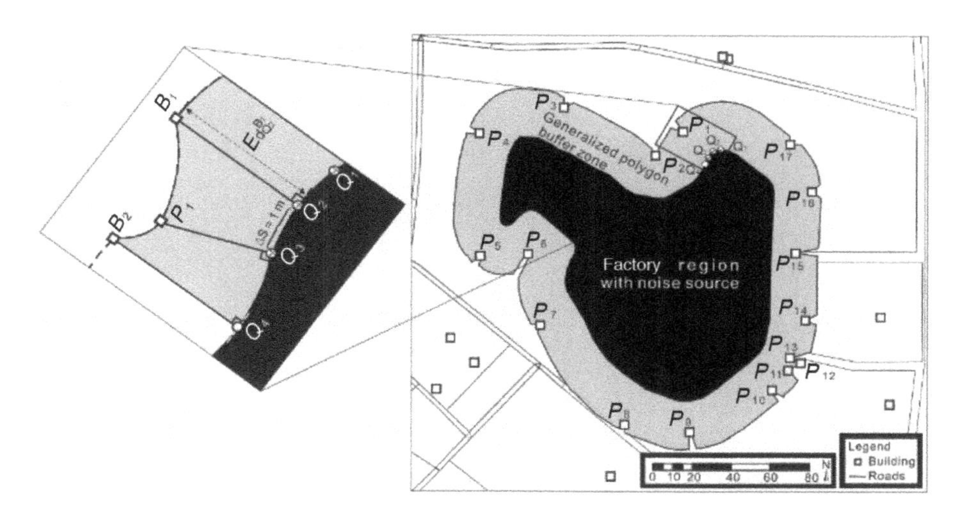

FIGURE 8.14 Generalized polygon buffer using Data set-1 with the increment curve at $\Delta S = 1.0$ m.

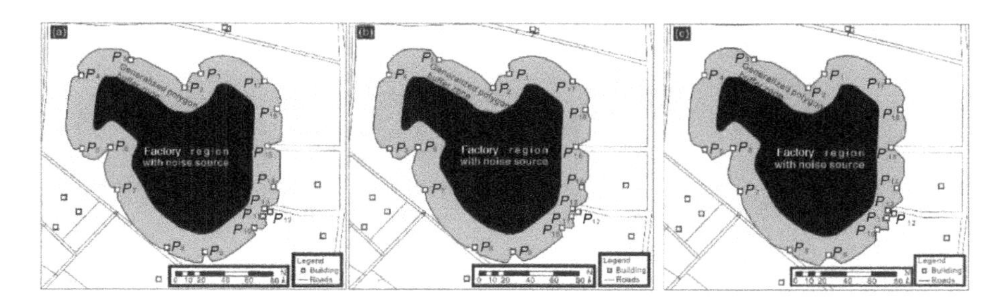

FIGURE 8.15 Generalized polygon buffer using Data set-1 with the different increment curves at (a) $\Delta S = 5.0$ m, (b) $\Delta S = 10.0$ m, and (c) $\Delta S = 15.0$ m.

same operations are repeated. The results are shown in Figure 8.15a, Figure 8.15b, and Figure 8.15c.

8.4.2.2 Experiment for Data Set-2

8.4.2.2.1 Generalized Point Buffer Zone Generation

For Data set-2, the same assumption of noise resource with 80 dB happens at Point *A*. The same operations are made as one for Data set-1, as described in Chapter 8.3.2. The traditional and generalized point buffer zone are generated. The results are depicted in Figure 8.16 and Figure 8.17, respectively.

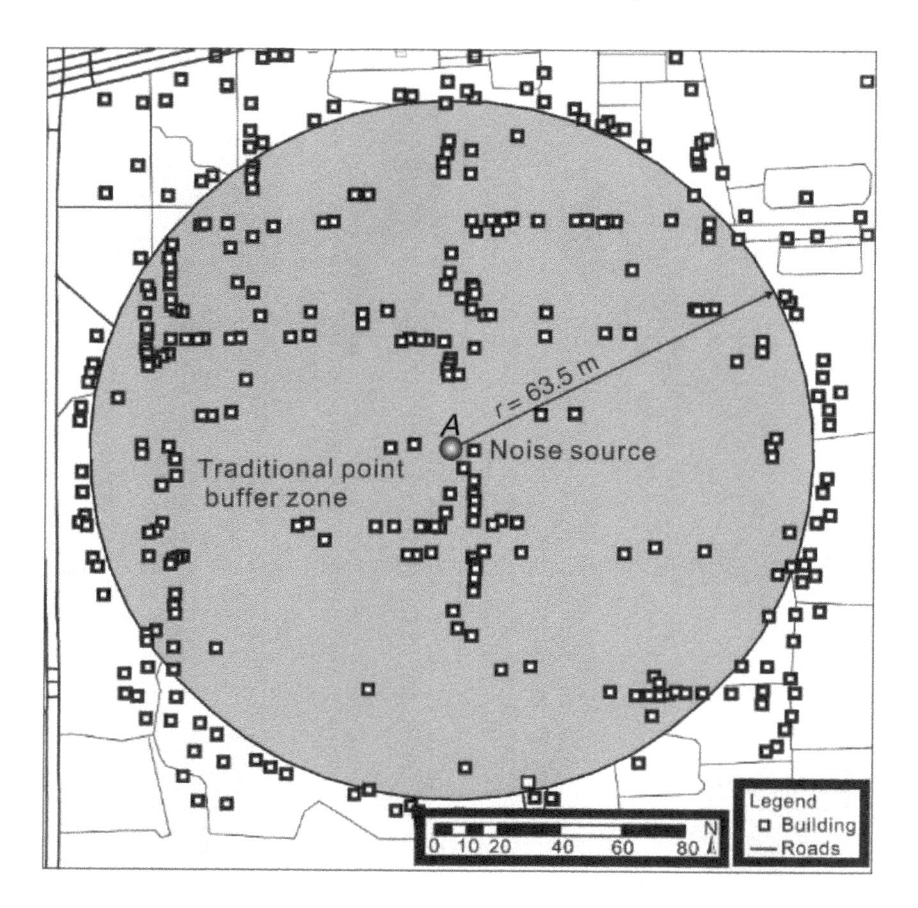

FIGURE 8.16 Traditional point buffer analysis with a radius of 63.5 m.

8.4.2.2.2 Generalized Line Buffer Analysis

Similarly, a vehicle's noise along the highway is considered as a noise source. The same operations are made as for Data set-1, as described in Chapter 8.4.1. The experimental results are shown in Figure 8.18 and Figure 8.19, respectively.

8.4.2.2.3 Generalized Polygon Buffer Analysis

Similarly, a noise at a factory is considered as a noise source. The same operations are made as the one for Data set-1, as described in Chapter 8.4.1. The experimental results are shown in Figure 8.20 and Figure 8.21.

8.4.3 Comparison Analysis and Remarks

8.4.3.1 Comparison Analysis

In order to compare the difference between the traditional and generalized buffer algorithms, the following parameters are used for indexes.

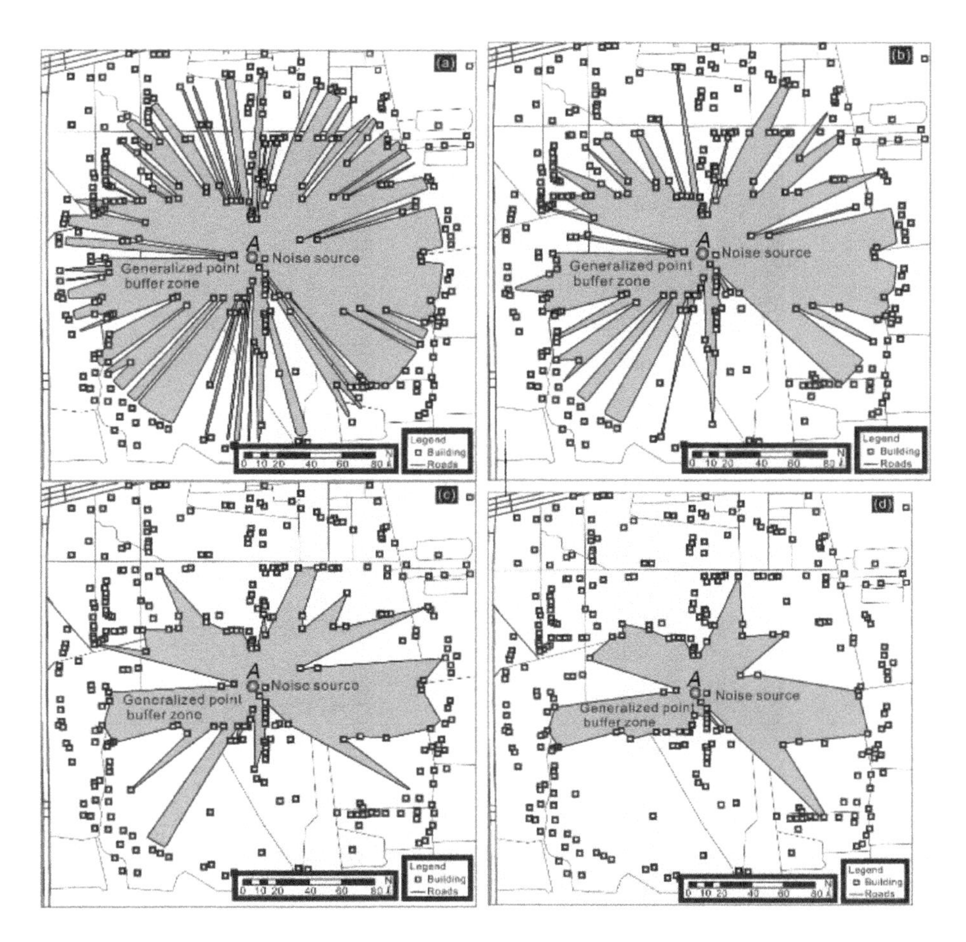

FIGURE 8.17 Generalized point buffer using Data set-2 with the different increment angles at (a) $\Delta\varphi_1 = 1.5°$, (b) $\Delta\varphi_2 = 3°$, (19c) $\Delta\varphi_3 = 6°$, and (d) $\Delta\varphi_4 = 9°$.

Area difference, which is the difference of the areas between the traditional buffer zone and generalized buffer zone, that is,

$$\Delta_{Area} = Area_T - Area_{G_i} \tag{8.25}$$

where Δ_{Area} is the difference of areas (m^2); $Area_T$ is the area from traditional buffer zone; $Area_{Gi}$ is area from the generalized point/line/polygon buffer zone; $G_i = Point$, *line*, and *Polygon*.

Perimeter difference, which is the difference of the perimeter between the traditional buffer zone and generalized buffer zone, that is,

$$\Delta_{Perimeter} = Perimeter_T - Perimeter_{G_i} \tag{8.26}$$

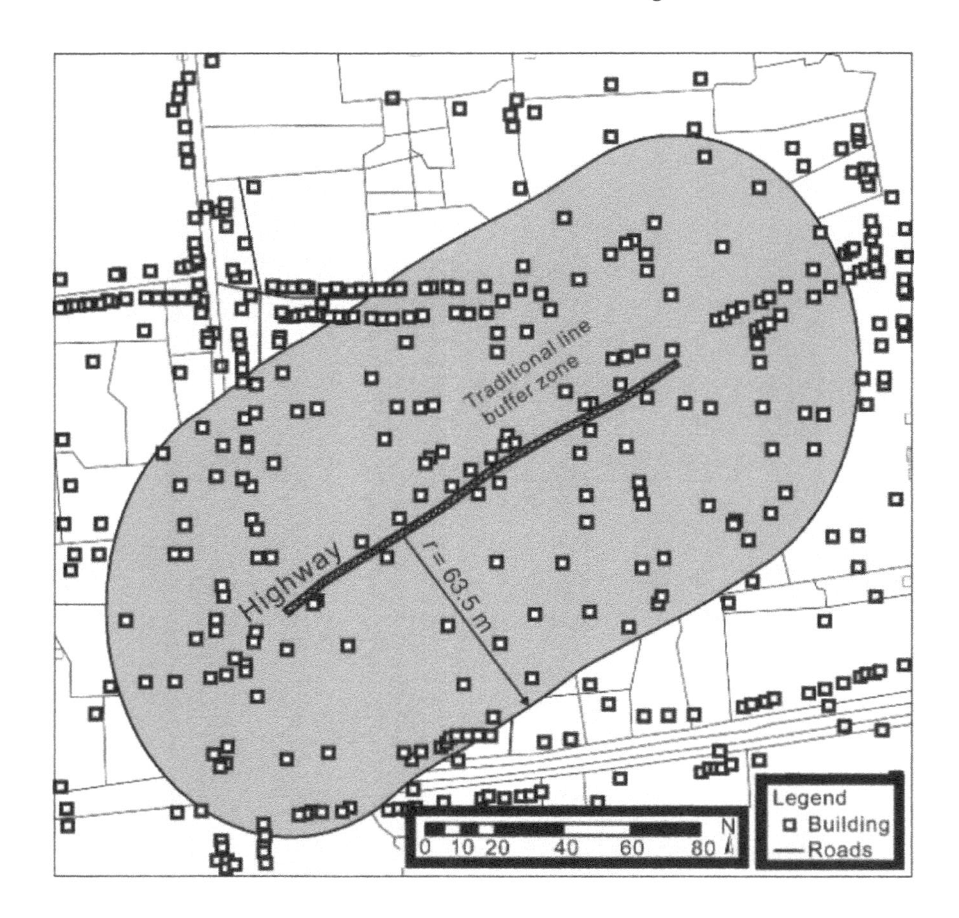

FIGURE 8.18 Traditional line buffer analysis with a distance of 63.5 m.

where $\Delta_{Perimeter}$ is the difference of perimeters (m); $Perimeter_T$ is the perimeter from the traditional buffer zone; $Perimeter_{Gi}$ is the perimeter from generalized point/line/polygon buffer zone.

Relative position similarity, which is defined as the distance between the centroids of the two buffers (Zhang et al. 2018), that is,

$$D(|a,b|) = \frac{1}{1+|a,b|^2} \tag{8.27}$$

where D(|a, b|) is the relative position and |*a*, *b*| is the distance between the two centroids.

Relative area similarity, which is defined as follows (Zhang et al. 2014):

$$S_{area} = 1 - \frac{|S_T - S_G|}{Max(S_T, S_G)} \tag{8.28}$$

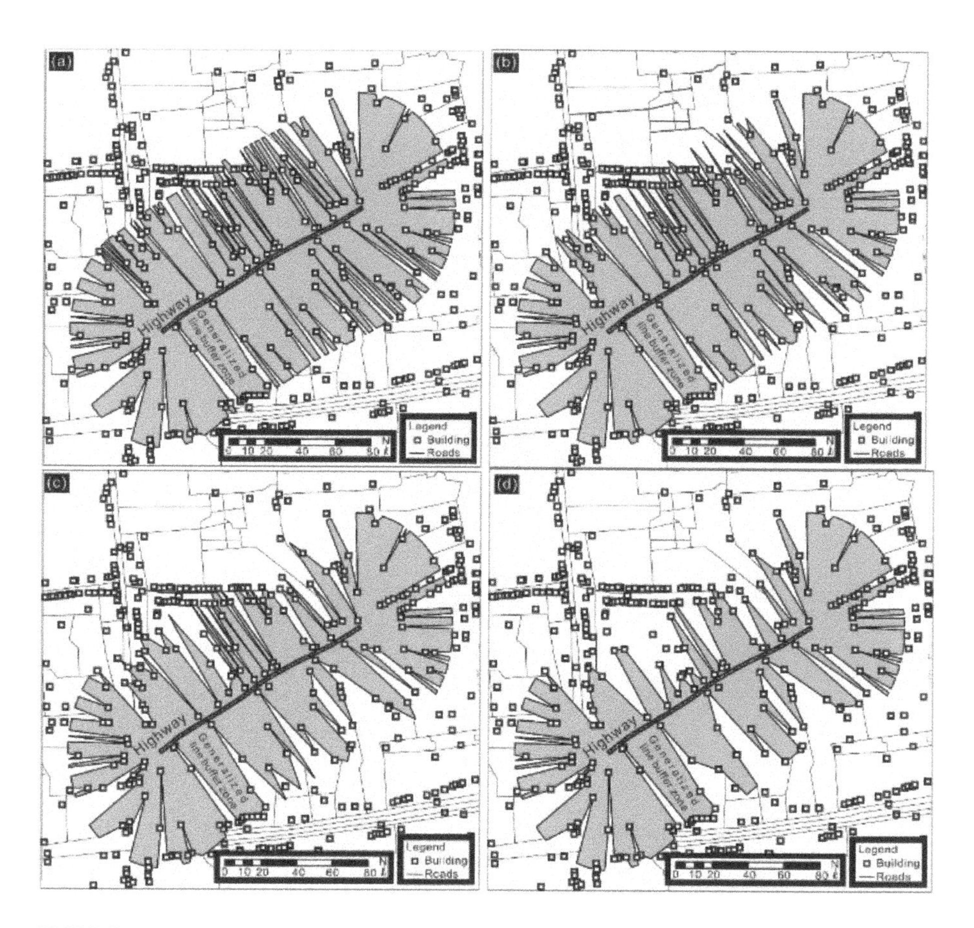

FIGURE 8.19 Generalized line buffer using Data set-2 with the different increment length at (a) $\Delta L_1 = 1.0$ m, (b) $\Delta L_2 = 5.0$ m, (c) $\Delta L_3 = 10.0$ m, and (d) $\Delta L_4 = 15.0$ m.

Where S_{area} is the relative area similarity, $|S_T - S_G|$ is the area difference of two buffer regions, and Max(S_T, S_G) is the maximum area of them.

Relative perimeter similarity, which is defined as follows (Zhang et al. 2014):

$$S_{per} = 1 - \frac{|P_T - P_G|}{Max(P_T, P_G)} \tag{8.29}$$

Where S_{per} is the relative perimeter similarity, $|P_T - P_G|$ is the difference between the perimeter of the traditional buffer and the generalized buffer, and Max(P_T, P_G) is the maximum perimeters of them.

Offset of centroid of mass, which is defined as

$$\Delta_f = \sqrt{\Delta X^2 + \Delta Y^2} \tag{8.30}$$

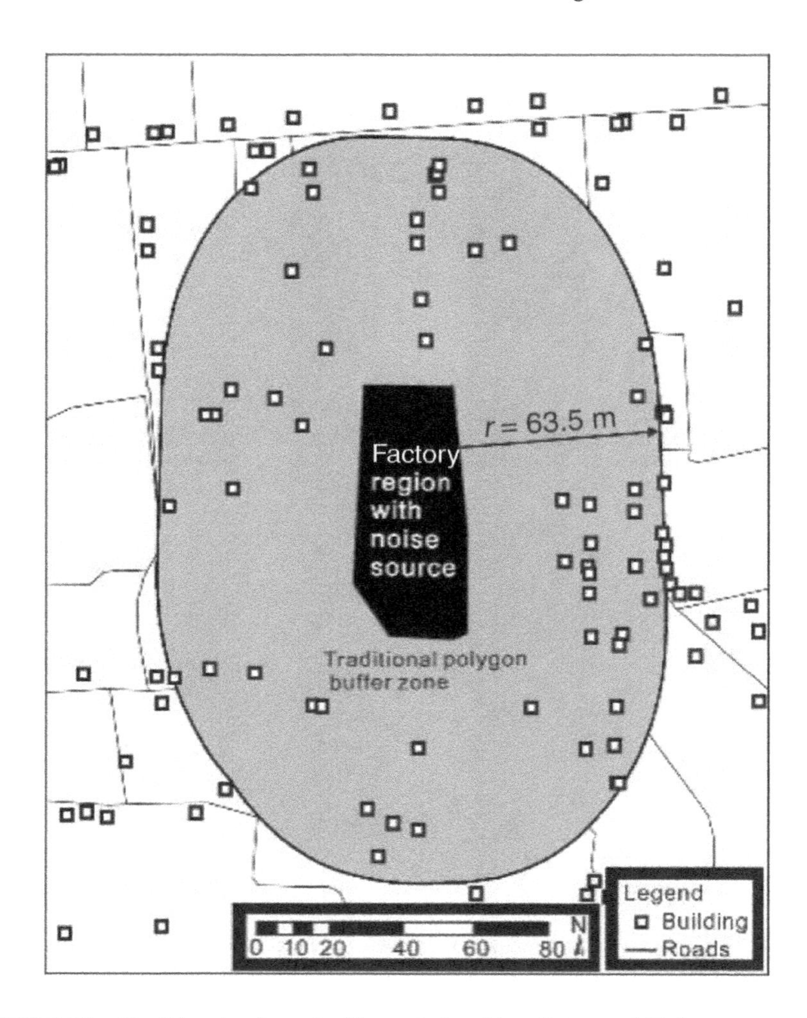

FIGURE 8.20 Traditional polygon buffer analysis with a distance of 63.5 m.

Where Δ_f is the offset of centroid mass, $\Delta X = X_T - X_{Gi}$ (i = 1,2,3,4), and is the difference between the X_T (X-coordinate of the centroid of the traditional buffer analysis) and the X_{Gi} (X-coordinate of centroid for generalized buffer analysis). $\Delta Y = Y_T - Y_{Gi}$ (i = 1,2,3,4) and is the difference between the Y_T (Y-coordinate of the centroid of the traditional buffer analysis) and the Y_{Gi} (Y-coordinate of centroid for generalized buffer analysis).

Comparison analysis for Data s et-1: For Data set-1, the six parameters for point, line, and polygon buffer zones are calculated, associated with the different parameters, and shown in Table 8.4 through Table 8.6.

Comparison analysis for Dataset-2: For Data set-2, the six parameters for point, line, and polygon buffer zones are calculated, associated with the different parameters, and shown in Table 8.6 through Table 8.7.

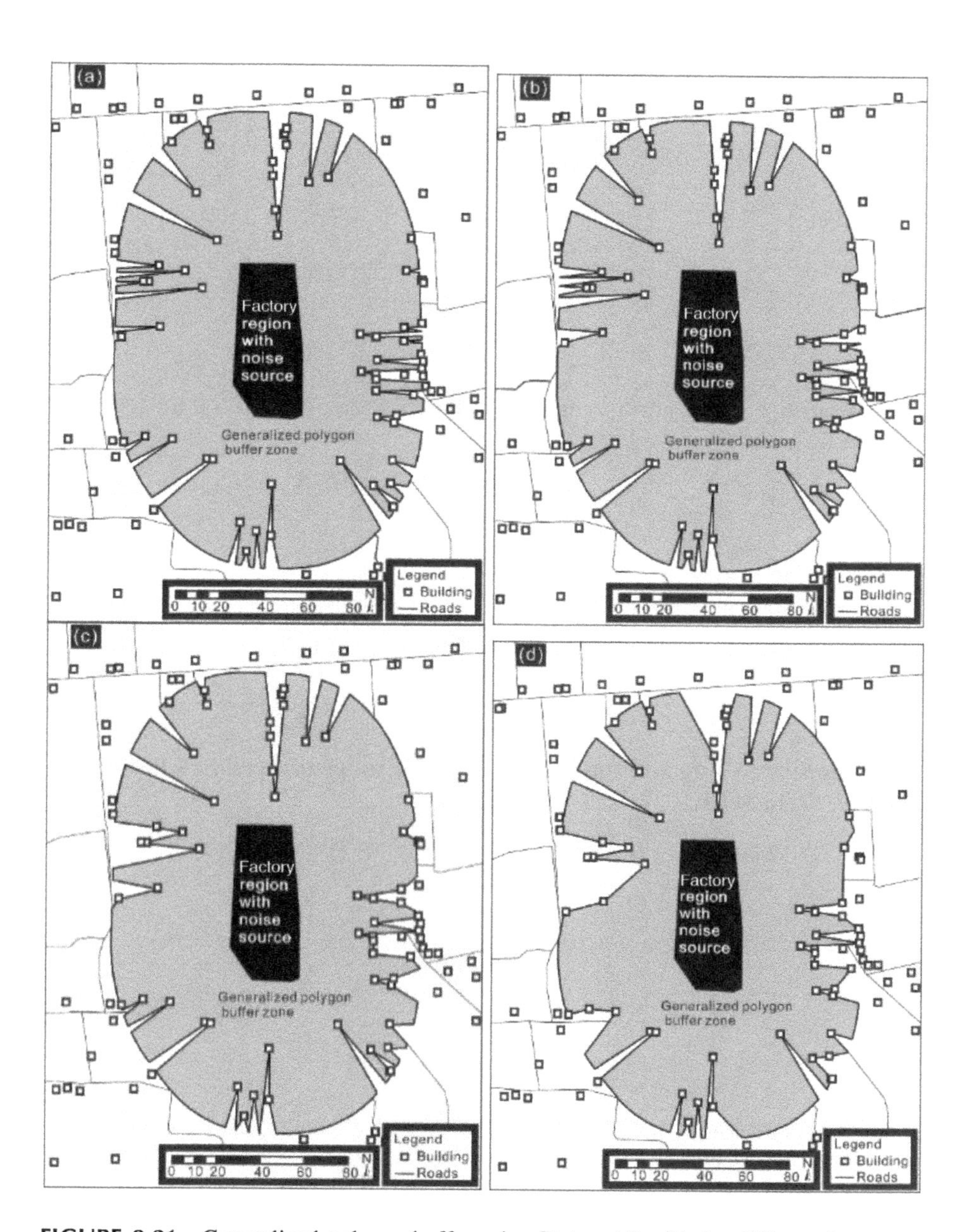

FIGURE 8.21 Generalized polygon buffer using Data set-2 with the different increment curves at (a)$\Delta S_1 = 1.0$ m, (b) $\Delta S_2 = 5.0$ m, (c) $\Delta S_3 = 10.0$ m, and (d) $\Delta S_4 = 15.0$ m.

TABLE 8.3
The Difference Value and the Shape Similarity Index of the *GPIB* Relative to *TPB* for Data Set-1

	Δ_{Area} (m^2)	$\Delta_{perimeter}$ (m)	Δ_f(m)	D (|a, b|) (%)	S_{area} (%)	S_{per} (%)
$\Delta\varphi_1 = 1.5°$	29	−205	1.4	41.4	97.7	66.1
$\Delta\varphi_2 = 3°$	46	−191	1.5	39.8	96.4	67.6
$\Delta\varphi_3 = 6°$	83	−167	2.2	31.2	93.5	70.5
$\Delta\varphi_4 = 9°$	112	−146	2.8	26.3	91.2	73.3

TABLE 8.4
The Difference Value and the Shape Similarity Index of the *GLB* Relative to *TLB* for Data Set-1

	Δ_{Area} (m^2)	$\Delta_{perimeter}$ (m)	Δ_f(m)	D (|a, b|) (%)	S_{area} (%)	S_{per} (%)
$\Delta L_1 = 1$ m	3697	−986	0.9	53.7	98.5	72.9
$\Delta L_2 = 5$ m	5897	−878	1.0	50.9	97.6	75.2
$\Delta L_3 = 10$ m	8805	−790	1.6	38.9	96.4	77.1
$\Delta L_4 = 15$ m	12286	−725	2.1	33.8	94.9	78.6

TABLE 8.5
The Difference Value and the Shape Similarity Index of the *GPLB* Relative to *TPLB* for Data Set-1

	Δ_{Area} (m^2)	$\Delta_{perimeter}$ (m)	Δ_f(m)	D (|a, b|) (%)	S_{area} (%)	S_{per} (%)
$\Delta S_1 = 1$ m	5034	−1425	17.3	5.8	97.7	68.3
$\Delta S_2 = 5$ m	8695	−1572	17.0	5.9	95.9	65.0
$\Delta S_3 = 10$ m	11796	−1652	16.9	5.9	94.6	63.2
$\Delta S_4 = 15$ m	16703	−1733	17.5	5.7	92.3	61.4

TABLE 8.6
The Difference Value and the Shape Similarity Indexes of the *GPIB* Relative to *TPB* for Data Set-2

	Δ_{Area} (m^2)	$\Delta_{perimeter}$ (m)	Δ_f(m)	D (|a, b|) (%)	S_{area} (%)	S_{per} (%)
$\Delta\varphi_1 = 1.5°$	394	−3565	34.1	2.8	68.9	10.1
$\Delta\varphi_2 = 3°$	690	−1502	47.6	2.1	45.6	20.9
$\Delta\varphi_3 = 6°$	867	−494	44.5	2.0	31.5	44.7
$\Delta\varphi_4 = 9°$	974	−84	52.6	1.9	23.1	82.6

TABLE 8.7
The Difference Value and the Shape Similarity Index of the *GLB* Relative to *TLB* for Data Set-2

	Δ_{Area} (m²)	$\Delta_{perimeter}$ (m)	Δ_f(m)	D (|a, b|) (%)	S_{area} (%)	S_{per} (%)
$\Delta L_1 = 1$ m	16050	−12493	8.5	10.6	80.3	11.1
$\Delta L_2 = 5$ m	24757	−11085	11.3	8.1	69.6	12.3
$\Delta L_3 = 10$ m	26349	−8741	14.7	6.4	67.6	15.1
$\Delta L_4 = 15$ m	47936	−3446	26.1	3.7	41.0	31.1

TABLE 8.8
The Difference Value and the Shape Similarity Index of the *GPLB* Relative to *TPLB* for Data Set-2

	Δ_{Area} (m²)	$\Delta_{perimeter}$ (m)	Δ_f(m)	D (|a, b|) (%)	S_{area} (%)	S_{per} (%)
$\Delta S_1 = 1$ m	7253	−2471	1.5	41.7	91.9	32.7
$\Delta S_2 = 5$ m	8539	−2372	1.4	41.3	90.6	33.2
$\Delta S_3 = 10$ m	10099	−2035	1.7	37.3	88.9	36.7
$\Delta S_4 = 15$ m	32034	−1792	2.0	33.7	64.7	39.6

8.4.3.2 Remarks from the Compared Results

8.4.3.2.1 From Traditional/Generalized Point Buffer Algorithms

With Table 8.4 and Table 8.7, a few remarks can be drawn up:

a. If the angle increment $\Delta\varphi$ limits to 0, the area of *GPIB* is close to the area of *TPB*, that is,

$$s_{GPIB} = \lim_{\Delta\varphi \to 0} \int_0^{2\pi} ds = s_{TPB} \tag{8.31}$$

This means that the generalized buffer zone is close to the traditional buffer zone when the angle increment $\Delta\varphi$ limits to 0.

b. Increasing the incremental angles ($\Delta\varphi$) from 1.5° to 9°, the area from GPB algorithm decreases from 1.6 times to 3.9 times for Data set-1 and from 1.8 times to 2.5 times for Data set-2 relative to ones from the TGB algorithm; the perimeters from the GPB algorithm dramatically increase from 0.7 times to 0.9 times for Data set-1 and from 0.02 times to 0.4 times for Data set-2. The offsets of mass centroid move from 1.4 m to 2.8 m for Data set-1 and from 34.1 m to 52.6 m for Data set-2.

c. Increasing the incremental angle ($\Delta\varphi$) from 1.5° to 9°, the relative position similarities decrease from 41.4% to 26.3% for Data set-1 and 2.8% to 1.9% for Data set-2. The relative area similarities decrease from 97.7% to 91.2% for

Data set-1 and 68.9% to 23.1% for Data set-2. The relative shape similarities increase from 66.1% to 73.3% for Data set-1 and 10.1% to 82.6% for Data set-2. This means that the relative area similarity, relative perimeter similarity, and the relative position similarity slightly increase for Data set-1, while the relative position similarity greatly increases for Data set-2.

8.4.3.2.2 *From Traditional/Generalized Line Buffer Algorithms*

With Table 8.5 and Table 8.8, a few remarks can be drawn up:

a. If the distance increment ΔL limit to 0, the area of *GLB* zone is close to the area of *TLB* zone, that is,

$$s_{GLB} = \lim_{\Delta L \to 0} \int_0^L ds = s_{TLB} \tag{8.32}$$

This means that the generalized line buffer zone is close to the traditional buffer zone when the line increment ΔL limits to 0.

b. Increasing the incremental distance (ΔL) from 1 m to 15 m, the areas of *GLB* zone decrease from 1.6 times to 3.3 times for Data set-1 and from 1.5 times to 3 times for Data set-2; the perimeters increase from 0.7 times to 0.9 times for Data set-1 and from 0.3 times to 0.9 times for Data set-2. The offsets of the mass centroid move from 0.9 m to 2.1 m for Data set-1 and from 8.5 m to 26.1 m for Data set-2.
c. Increasing the incremental distance (ΔL) from 1 m to 15 m, the relative position similarities decrease from 53.7% to 33.8% for Data set-1 and 10.6% to 3.7% for Data set-2. The relative area similarities decrease from 98.5% to 94.9% for Data set-1 and 80.3% to 41.0% for Data set-2. The relative perimeter similarities with increasing the incremental distance increase from 72.9% to 78.6% for Data set-1 and 11.1% to 31.1% or Data set-2. This means that the relative area similarity, relative perimeter similarity, and the relative position similarity slightly increase for Data set-1, while the relative position similarity greatly increase for Data set-2.

8.4.3.2.3 *From Traditional/Generalized Polygon Buffer Algorithms*

With Table 8.6 and Table 8.8, a few remarks can be drawn up:

a. If the arc increment ΔS approaches 0, the area of *GPLB* zone is close to the area of the *TPLB* zone, that is,

$$S_{GPLB} = \lim_{\Delta S \to 0} \int_0^S ds = S_{TPLB} \tag{8.33}$$

This means that the generalized polygon buffer zone is close to the traditional polygon buffer zone when the arc increment ΔS limits to 0.

b. Increasing the incremental arc length (ΔS) from 1m to 15 m, the areas of *GPLB* decrease from 1.7 times to 3.3 times for Data set-1 and from 1.2 times to 4.4 times for Data set-2, and the perimeters increase from 1.1 times to 1.2 times for Data set-1 and from 0.7 times to 0.9 times for Data set-2. The offsets of mass centroid move from 17.3 m to 17.5 m for Data set-1 and from 1.4 m to 2.0 m for Data set-2.

c. Increasing the incremental arc length (ΔS) from 1 m to 15 m, the relative position similarities decrease from 5.8% to 5.7% for Data set-1 and 41.7% to 33.7% for Data set-2. The relative area similarities decrease from 97.7% to 92.3% for Data set-1 and 91.9% to 64.7% for Data set-2. The relative perimeter similarities increase from 68.3% to 61.4% for Data set-1 and 32.7% to 39.6% for Data set-2. This means that the relative area similarity and relative perimeter similarity slightly increase, while the relative position similarity quickly increase for Data set-1, and the relative area similarity, relative perimeter similarity, and relative position similarity quickly increase for Data set-2.

With this analysis, the following conclusions can be drawn up:

a. The differences between traditional and generalized buffer zones for the point, line, and polygon buffer zone generation are different upon the variables, such as incremental angles ($\Delta\varphi$), incremental length(ΔL), and incremental arc length (ΔS). In particular, when these increments approach 0, the traditional and generalized buffer zones are the same.
b. Increasing the incremental angle ($\Delta\varphi$), incremental length (ΔL), and incremental arc length (ΔS), the relative position similarity in GB zones for the point, line, and polygon buffering zone generation decreases. The smaller the data density is, the bigger the relative position similarity changes, and vice versa. The higher the relative position similarity is, the simpler the generalized buffer shape is, even approaching the traditional buffer shape, and vice versa.
c. Increasing the incremental angle ($\Delta\varphi$), incremental length (ΔL), and incremental arc length (ΔS), the relative area similarity of the generalized buffering zone decreases. The smaller data density is, the smaller relative area similarity changes, and vice versa. The higher the relative area similarity is, the closer the generalized buffer shape is to the traditional buffer shaper, and vice versa.
d. Increasing the incremental angle ($\Delta\varphi$), incremental length (ΔL), and incremental arc length (ΔS), the perimeters for point, line, and polygon buffer zones increase. In addition, the data density is small, and the change in the relative perimeter similarity is small. When the data density is large, the relative perimeter similarity changes greatly. The higher the relative perimeter similarity the simpler the generalized buffer shape is, and the closer the generalized buffer shape is to the traditional buffer.

8.5 CONCLUSION

The main contribution of this research is the development of a radial and breakthrough algorithm called a *generalized buffer algorithm* (*GBA*) for point, line, and polygon buffer generation. This algorithm is challenging the traditional buffering algorithm that has been used for over 60 years. This algorithm stems from the fact that traditional buffering algorithms are based on a fixed buffer distance without considering the difference of neighbor instances' attributes in practice. The proposed *GBA* simultaneously considers homogeneity and the correlation of both spatial data and attribute data of two instances; consequently, the buffer distance varies upon the characteristics of two instances.

The details of the proposed *GBA* are described in text. In summary, first, spatial and nonspatial attributes are selected. Second, the *R*-proximity relationships between two instances are determined in accordance with the selected spatial attributes. Third, candidates of boundary points of the buffering zone are selected based on the *R-proximity* relationship. Forth, boundary points of the buffering zone are determined using nonspatial attributes to decide if the candidates of boundary points of the buffer zone are prevalent events. Finally, the boundary points are connected to form the boundary of the generalized buffer zone. To validate the advances of the proposed method, the point, line, and polygon data sets from Beijing and Bao'an District, City of Shenzhen, China, are used. The experimental results and comparison analyses, using six indexes calculated from traditional and generalized buffer algorithms, discovered that:

1. The proposed GBA can accurately reflect the real situation of the buffering zone and improve the deficiency and accuracy of the traditional buffering algorithm.
2. From six indexes, GBA approaches to the traditional point/line/polygon buffering algorithms when the incremental angle ($\Delta\varphi$), the incremental length (ΔL), and the incremental arc length (ΔS) approach zero.

REFERENCES

Aggarwal, A., Chazelle, B., Guibas, L., et al., Parallel computational geometry. *Presented at IEEE Symposium on Foundations of Computer Science*, 1985. https://link.springer.com/article/10.1007/BF01762120.

Bai, F., and Zhang, F., Multi-objective buffer generation algorithms. *GNSS World of China*, vol. 36, no. 1, 2011, pp. 38–41.

Chen, J., Zhao, R., and Li, Z., Voronoi-based k-order neighbour relations for spatial analysis. *ISPRS Journal of Photogrammetry and Remote Sensing*, vol. 59, no. 1–2, 2004, pp. 60–72.

Chen, N., Influence of conversion on the location of points and lines: The change of location entropy and the probability of a vector point inside the converted grid point. *ISPRS Journal of Photogrammetry & Remote Sensing*, vol. 137, 2018, pp. 84–96.

Chen, Q., and Liu, G., Algorithm for generating irregular buffer zone with constraints. *Geological Science and Technology Information*, vol. 4, 2014, pp. 213–218.

Chvátal, V., On certain polytopes associated with graphs. *Journal of Combinatorial Theory*, vol. 18, no. 2, 1975, pp. 138–154.

Dong, P., et al., An effective buffer generation method in GIS. *Presented at IEEE IGARSS*, August 2003.

Dong, Q., and Jing, C., A tile-based method for geodesic buffer generation in a virtual globe. *International Journal of Geographical Information Science*, vol. 32, no. 2, 2017, pp. 302–323.

Er, E., Ismail, K., Görkem, G., et al., A buffer zone computation algorithm for corridor rendering in GIS. *IEEE International Symposium on Computer & Information Sciences*, September 2009. [Online]. https://ieeexplore.ieee.org/document/5291855.

Fan, J., et al., Optimization approaches to mpi and area merging-based parallel buffer algorithm. *Boletim de Ciências Geodésicas*, vol. 20, 2014, pp. 237–256.

Huang, S.X., Buffer analysis method based on spatial co-location pattern mining algorithm. *MS. Thesis*, Guilin University of Technology, Guilin, China, July 2020.

Lee, J., and Who, C., Topology preserving relaxation labeling for nonrigid point matching. *IEEE Transactions on Pattern Analysis & Machine Intelligence*, vol. 33, no. 2, 2011, pp. 427–432.

Liu, L., and Min, J., Analysis of traditional space buffer generation algorithm. *Computer CD Software and Application*, vol. 35, no. 20, 2011, p. 35.

Ma, M., et al., Interactive and online buffer-overlay analytics of large-scale spatial data. *International Journal of Geo-Information*, vol. 8, no. 1, 2019, pp. 21–34.

Ma, X., and Chang, G., The influence of vehicle sound source height on traffic noise forecast. *Science Technology and Engineering*, vol. 36, 2018, pp. 180–184.

Pan, W., and Li, Y., Random algorithm for buffer generation. *Computer Engineering*, vol. 36, no. 14, 2010, pp. 70–73.

Preparata, F.P., and Shamos, M.I., *Computational geometry: An introduction*. Springer-Verlag, Berlin, Germany, 1985.

Ren, Y., et al., A way to speed up buffer generalization by Douglas-Peucker algorithm. *Presented at IEEE International Geoscience & Remote Sensing Symposium*, September 2004. [Online]. https://ieeexplore.ieee.org/document/1370304.

Shimrat, M., Algorithm 112: Position of point relative to polygon. *Communications of the ACM*, vol. 5, no. 8, 1962, p. 434.

Wang, E., et al., Edge match algorithm based on attributes similarity and multiple buffers. *Bulletin of Science and Technology*, vol. 32, no. 4, 2016, pp. 174–177.

Wang, T., et al., Parallel research and optimization of buffer algorithm based on equivalent arc partition. *Remote Sensing Information*, vol. 31, no. 4, 2016, pp. 147–152.

Wu, H., On the establishment of GIS buffer zone. *Journal of Wuhan University of Surveying and Mapping*, vol. 22, no. 4, 1997, pp. 358–366.

Wu, H., Gong, J., and Li, D., Buffer curve and buffer generation algorithm in aid of edge-constrained triangle network. *ACTA Geodaetica et Cartographica Sinica*, vol. 4, 1999, pp. 75–79.

Xu, X., and Liu, W., Research on vector-raster mixed algorithm of linear buffer generation. *Computer Engineering and Applications*, vol. 50, no. 4, 2014, pp. 152–156.

Yang, C., et al., Algorithm for rapid buffer analysis in three-dimensional digital earth. *Journal of Remote Sensing*, vol. 18, no. 2, 2014, pp. 353–364.

Zhang, X., Ai, T., and Stoter, J., Data matching of building polygons at multiple map scales improved by contextual information and relaxation. *ISPRS Journal of Photogrammetry and Remote Sensing*, vol. 92, 2014, pp. 147–163.

9 Application of Mining Co-Location Patterns in Remotely Sensed Imagery Classification

9.1 INTRODUCTION

In this section, we describe the test data Set and experiment environment and present the results of land cover classification. In addition to the proposed method, SAE (a deep neural network) that is regarded as baseline, CL-DT, and traditional DT (CART) are also used to classify the land cover of two test areas. The details are described as follows.

9.2 DATA SETS

9.2.1 DATA SETS

1. The first test area and data set: The first test area is located at 24.25° thru 26.38° north latitude and 109.60° thru 111.48° east longitude, covering the entire city of Guilin, Guangxi Province, China. The area covers approximately 27,200 km^2. The test area is a typical karst plain landform in which there are many exposed carbonate rocks (Zhou et al. 2010).

The images of Landsat-5 TM were acquired at a local time on September 21, 2006. The spatial resolution of three visible bands, one near-infrared band, one middle infrared band, and one far infrared are 30 meters, and the spatial resolution of one thermal infrared band is 120 meters. The Landsat-5, with the main detector TM, is operated in a sun-synchronous, 705 km orbit height and a 16-day revisit cycle.

2. The second test area and data set: The second test area is located at 23.78° through 24.58° north latitude and 107.85° thru 108.5° east longitude, covering Du'an County of the city of Hechi, Guangxi Province, China. The area covers nearly 4,459 km^2. The test area is a typical karst rocky desertification area in which exposed carbonate rocks are widespread.

The images of Landsat-5 TM were acquired at local time on January 30, 2009, and were ordered from the website http://datamirror.csdb.cn/.

These imagery data were preprocessed before further utilization, including geometric correction (Kardoulas et al. 1996; Storey and Choate 2004; Zhou 2011; Zhou and Wang 2012.), mosaic (Yang 1990; Kanazawaa and Kanatani 2004), and clipping (Greiner and Hormann 1998).

DOI: 10.1201/9781003139416-9

9.2.2 Nonspatial Attribute and Spatial Attribute Selection

In addition to the original TM imagery data, as a rule of thumb, five nonspatial attribute data, including SSM, LST, VC, the components of PCA, and texture (TEX), are considered to classify the land cover (see Table 9.1). SSM data can be generated by the method of the spectrum of soil water content and the regression method (Liu et al. 1997). LST is retrieved from Landsat TM data by a mono-window algorithm (Qin and Karniell 2001). VC can be obtained by the vegetation index method (Mohammad et al. 2002). By applying the PCA function of ENVI 4.8 software for the preprocessed TM imagery data, the components of PCA can be acquired, whose *i*-th component can be represented by PCA_i. Based on the co-occurrence measures, the texture (noted as TEX) data can be produced. There are 30,217,776 instances for each attribute in the first test area and 4,954,112 instances in the second test area.

As a rule of thumb, the instances are classified into five categories: water (WT), vegetation (VG), exposed carbonate (EC), habitation (HB), and cultivated land (CL).

In addition to these nonspatial attributes, spatial attributes including X and Y coordinates are considered as well. The metadata for the spatial data are listed in Table 9.2.

TABLE 9.1
Four Nonspatial Attributes

#	Nonspatial attribute	Scope	
1	The components of PCA	WT	$PCA_1 > 30$
		VG	$PCA_2 > 5$
		EC	$PCA_3 < -5$
		HB	$PCA_1 < -10$ and PCA3 > 10
		CL	$PCA_1 < -20$
2	VC	The scope is 0.1 to 0.9.	
3	SSM	Percentages of I ≤ 5%, 5% < II ≤ 10%, 10% < III ≤ 15%, 15% < IV ≤ 20%, 20 < V ≤ 25%, VI > 25, indicate different levels of moisture content.	
4	LST	The scope is 282K to 302K	

TABLE 9.2
X/Y Coordinates

#	Description	
X/Y coordinates	Projection: Transverse_Mercator; False_Easting: 500000.000000; False_Northing: 0.000000; Central_Meridian: 111.000000; Scale_Factor: 0.999600; Latitude_Of_Origin: 0.000000; Linear Unit: Meter (1.000000);	Geographic Coordinate System: GCS_WGS_1984 Angular Unit: Degree (0.017453292519943295) Prime Meridian: Greenwich (0.000000000000) Datum: D_WGS_1984 Spheroid: WGS_1984

Furthermore, these TM imagery data, SSM, LST, VC, the components of PCA, TEX, and X/Y coordinates are managed by a database, and each of them is respectively regarded as one of the dimensions of high dimensional space in which instances belong to nonlinear distribution. Therefore, the coordinate of every instance can be seen as a vector expressed by (att_{11}, att_{12}, . . ., att_{1n}), where att_{ij} represents the value of the i-th instance in the j-th attribute.

9.3 EXPERIMENTS

The flowchart of the experimental procedure is depicted in Figure 9.1. First, the MVU algorithm is utilized to "unfold" input data, and MVU unfolded distance between instances are calculated. Second, the unfolded distance is merged with the co-location mining algorithm to establish the exact *RRS* (r-relationship) between instances, and then MVU-based co-location rules are mined. Finally, the MVU-based co-location rules are used to induce the generation of the decision tree, and MVU-based co-location decision rules are obtained.

The experiment was conducted using a computer with an Intel Xeon E5645 Six core 2400 MHz (12 MB Cache) and 4 GB of RAM. We implement the proposed algorithm in ENVI+IDL 4.8. For all experiments, we have used tenfold cross validation. Through Google Earth, the regions of interest (ROIs) of five classes (WT, VG, EC, HB, and CL) are acquired in Landsat TM images (see Table 9.3), respectively. Tenfold cross-validation breaks data sets of ROIs into 10 subsets of size $N/10$. It trains the proposed classifier using nine subsets and tests it using the remaining one subset.

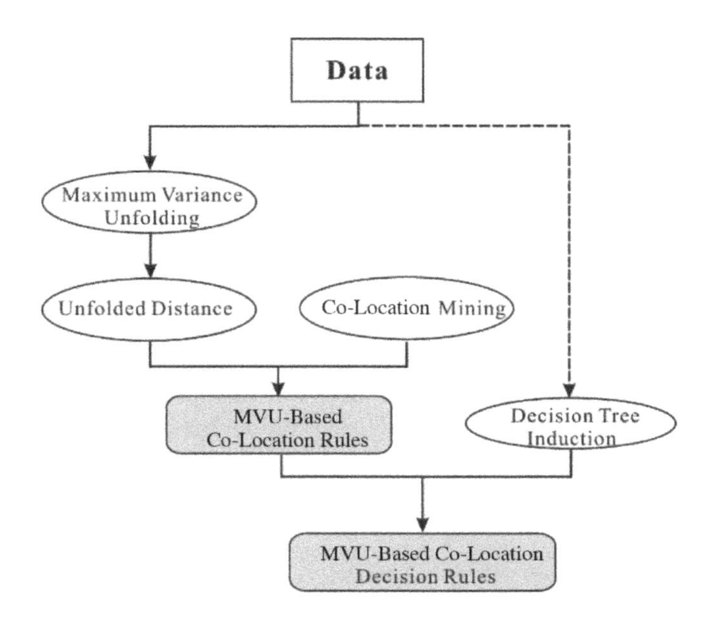

FIGURE 9.1 The flowchart of the experiment.

TABLE 9.3
Data Sets of ROIs of First Test Area

Data sets	#Instances	#Attributes
WT	10,000	12
VG	10,000	12
EC	10,000	12
HB	10,000	12
CL	10,000	12

9.3.1 Experiments on the First Test Area

9.3.1.1 Input Parameters

Data input: The MVU-based co-location database, which has completed the "preprocessing" using the MVU-based co-location mining method, is loaded. Additionally, a subset of data is applied to build the model, and the rest is utilized to validate the performance of the model.

Parameters input: It is necessary to input some parameters to optimize the process of DT induction.

These parameters include:

a. Minimum node size: The minimum valid node size is 0.
b. Maximum purity: If the purity of a node is 95% or higher, the algorithm stops splitting it. Additionally, if its number of records is 1% or less of the total number of records, the algorithm stops splitting it.
c. Maximum depth: The maximum valid depth is 20.

9.3.1.2 Generation of MVU-Based Co-Location Mining Rules

9.3.1.2.1 Calculation of Unfolded Distances

Generally, there is a high correlation among different bands of remotely sensed imagery. And most data are redundant and repeated from the viewpoint of extracting the useful information. The principle component analysis (PCA) can integrate the useful information of original multi-band images together and make these principal component images unrelated with each other. For example, as shown in Table 9.4, the correlation between an arbitrary two bands of Landsat-5 thematic mapper (TM) is high, especially the correlation between TM2 and TM3. After performing PCA, these principal component images are unrelated with each other (see Table 9.5), and the main information is contained in the former three components.

To save computational time of data processing in the MVU-based CL-DT generation, this chapter proposes an initial processing after eliminating the correlation between images using the PCA method to delete those instances that are not apparently neighbors. This chapter takes one attribute, PCA_1, which represents the

TABLE 9.4
The Correlation between an Arbitrary Two Bands of Landsat-5

#	TM1	TM2	TM3	TM4	TM5	TM7
TM1	1.0	0.95	0.92	0.31	0.46	0.62
TM2		1.0	0.98	0.39	0.58	0.72
TM3			1.0	0.35	0.63	0.79
TM4				1.0	0.68	0.45
TM5					1.0	0.91
TM7						1.0

TABLE 9.5
The Correlation between an Arbitrary Two Components of PCA

#	PCA_1	PCA_2	PCA_3	PCA_4	PCA_5	PCA_6
PCA_1	1.0	0	0	0	0	0
PCA_2		1.0	0	0	0	0
PCA_3			1.0	0	0	0
PCA_4				1.0	0	0
PCA_5					1.0	0
PCA_6						1.0

first components of PCA in image processing, as an example to explain the initial processing.

Different thresholds are set up for the instances of different categories to determinate the initial candidates of the MVU-based co-location. For example, the thresholds of PCA_1 are set as $r_{\theta 1}$ and $r_{\theta 2}$ ($r_{\theta 1} < r_{\theta 2}$) for the exposed carbonate. If the values of PCA_1 are within $r_{\theta 1}$ and $r_{\theta 2}$, the instances with eligible values are the initial candidate instances. It can be mathematically expressed by

$$PCA_1(i,j) = \begin{cases} PCA_1(i,j) & \text{if } r_{\theta 1} \leq PCA_1(i,j) \leq r_{\theta 2} \\ 0 & \text{other} \end{cases} \quad PCA_1(i,j) \subseteq P \tag{9.1}$$

where $r_{\theta 1}$ and $r_{\theta 2}$ are the given minimum and maximum thresholds of PCA_1, respectively; P is the set of $PCA_1(i, j)$; and i and j represent the i-th row and j-th column, respectively.

With the implementation of this step, those non-neighbored instances will be deleted, and the other instances are called *initial candidate instances*. This means that the non-neighbored instances are excluded in the following computation. As a result, computational time is saved.

From equation 9.1, the components of PCA are selected to determine the initial candidates of the MVU-based co-location. For example, the threshold of PCA_3, as for the exposed carbonate, is less than or equal to –5, that is,

$$\begin{cases} Max\{tr(L)\} \\ s.t.: \quad (1)\,\delta_{ik}(L_{ii} - 2L_{ik} + L_{kk}) = \delta_{ik} D_{ik} \\ \qquad (2) \sum_{ik} L_{ik} = 0 \\ \qquad (3)\, L \geq 0 \end{cases} \tag{9.2}$$

$$D_U(Y_i, Y_j) = \begin{cases} D_E(Y_i, Y_j) & \text{if } Y_i \text{ and } Y_j \text{ are neighbor} \\ min\{D_U(Y_i, Y_j), D_U(Y_i, Y_k) + D_U(Y_k, Y_h) + \cdots + D_U(Y_g, Y_j)\} & \text{others} \end{cases} \tag{9.3}$$

$$PCA_3(i, j) = \begin{cases} PCA_3(i, j) & \text{if } PCA_3(i, j) \leq -5 \\ 0 & \text{if } PCA_3(i, j) = other \end{cases} \quad PCA_3(i, j) \subseteq P \tag{9.4}$$

When the filtering is successfully finished above, the MVU algorithm is utilized to unfold these data using equation 9.2. After unfolding these data, the unfolded distances between instances are calculated using equation 9.3. After calculation, a sparse matrix $D_{m \times m}$ of unfolded distance can be obtained that records the unfolded distances among instances.

$$D_{m \times m} = \begin{Bmatrix} \begin{bmatrix} 0 & 30 & 60 & \cdots & 230025 \\ & 0 & 30 & \cdots & \vdots \\ & & \cdots & \cdots & \cdots \\ & & & \cdots & 30 \\ & & & & 0 \end{bmatrix}_{m \times m} \end{Bmatrix} \tag{9.5}$$

where m is 30,217,776, and the unit is meters.

9.3.1.2.2 Determination of the MVU-Based Co-Location

After obtaining the unfolded distances between instances, the *RRS*s between instances are determined using equation 9.6. In equation 9.6, the threshold of unfolded distance is set as 60, that is, $D_\theta = 60$ (meters) because the resolution of the TM images is 30 meters. Therefore, we have

$$R(Y_i, Y_j) = \begin{cases} 1 & \text{if } D_U(Y_i, Y_j) \leq 60 \\ \text{NAN} & \text{if } D_U(Y_i, Y_j) > 60 \end{cases} \tag{9.6}$$

With this determination of the *RRS*, those instances with *RRS* are the candidates for the MVU-based co-location. A sparse matrix $CD_{m \times m}$ of the candidates for MVU-based co-location is generated.

$$DR = \frac{similar\_AttrN(Y_i, Y_j)}{Total\_AttrN} \tag{9.7}$$

$$CD_{m \times m} = \begin{Bmatrix} \begin{bmatrix} 0 & 1 & 1 & \cdots & 0 \\ & 0 & 1 & \cdots & \vdots \\ & & \cdots & \cdots & \cdots \\ & & & \cdots & 1 \\ & & & & 0 \end{bmatrix}_{m \times m} \end{Bmatrix} \tag{9.8}$$

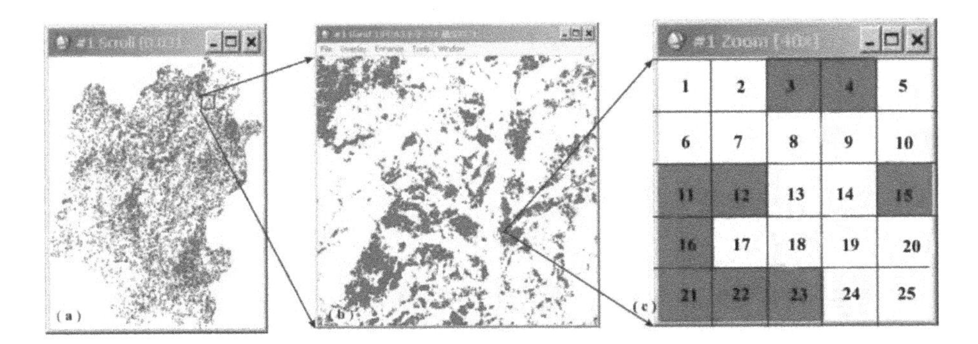

FIGURE 9.2 (a) The results of the determination of table instances; (b) the results are amplified two times; (c) the results are amplified 40 times.

TABLE 9.6
Determination of Table Instances

Pattern	*RRS*	PCA3	VC	LST	SSM (%)	TEX	DR
(2,3)	Yes	(−20.4,−22.3)	(0.33,0.15)	(17.2,16.8)	(4.8,6.8)	(53.7,46.3)	2/5
(3,4)	Yes	(−22.3,−18.9)	(0.15,0.13)	(16.8,16.3)	(6.8,7.6)	(46.3,45.7)	1
(10,15)	Yes	(−17.2,−20.8)	(0.21,0.24)	(16.9,16.8)	(9.8,5.3)	(58.7,46.5)	3/5
(11,12,16)	Yes	(−19.3,−19.8,−21.2)	(0.21,0.19,0.24)	(16.5,17.1,16.6)	(6.1,7.4,6.7)	(47.6,46.6,6.9)	1

On the basis of equation 9.7 and equation 9.8, for each pair candidate of MVU-based co-location instances, PCA components, VC, SSM, LST, and TEX are employed to determine the MVU-based co-locations. Taking the EC as an example, as mentioned earlier, if the *DR* (density ratio) of candidates of MVU-based co-location of the exposed carbonate is greater than or equal to 4/5, they are MVU-based co-locations. If not, they will be deleted from the candidates. With the application of this algorithm, the *k*-th order MVU-based co-locations can be obtained, as shown in Figure 9.2.

With this calculation, 8,122,155 instances are preserved. Figure 9.2c only shows 25 instances, and the 1st-order (ignored) through 6th-order table instances of MVU-based co-location are presented. Table VI takes four examples to explain the generation processes of the determination of MVU-based co-location. Figure 9.6c only shows the *DR*s of candidate patterns (3, 4) and (11, 12, 16), which are greater than 4/5. However, for patterns (2, 3), there is an *RRS* between them, the TEX is greater than 50, the SSM is not in the scope of 5 to 8, and the VC is not less than 0.25 (and similarly, for pattern (10, 15)), so they are deleted from the candidate set.

9.3.1.2.3 Determination of Distinct-Type Events

On the basis of equation 9.8, for EC, the cluster center value of attributes (such as VC) can be obtained from the average value of samples. So we have

$$\Psi_i = \sum_{i=1}^{S}\sum_{k=1}^{C}(\|f_i - m_k\|)^2 \tag{9.8}$$

$$\Psi_i = (\|f_i - 0.37\|)^2 \tag{9.9}$$

With the calculation, the threshold is set to 0.04. If Ψ_i is less than or equal to 0.04, then the i-th instance is a distinct event.

9.3.1.3 Experimental Results

A DT is induced by the proposed method. With post-processing to the DT, five decision rules are induced (Figure 9.3).

The classification results of remote sensing images in the first test area using the proposed method are presented in Figure 9.4a. Moreover, through applying CL-DT, SAE (Chen et al. 2014; Vincent et al. 2010), and traditional DT (CART), the classification results of remote sensing images in the first test area are depicted in Figure 9.4b, Figure 9.4c, and Figure 9.4d, respectively. Furthermore, some differences are marked by black circles in Figure 9.4. In addition, the distribution of exposed carbonate is shown in Figure 9.5.

9.3.2 Experiments on the Second Test Area

Similarly, the classification of remote sensing images in the second test area is retrieved by using the same method as in the first test area. The classification results in the second test area using the proposed method, CL-DT, SAE, and DT are shown in Figure 9.6a, Figure 9.6b, Figure 9.6c, and Figure 9.6d, respectively. Furthermore, some differences are marked black circles in Figure 9.6. In addition, the distribution of exposed carbonate in second test area is shown in Figure 9.7.

Rule 1:
If ("vegetation coverage">0.66 and "PCA2>5") Then "Vegetation";

Rule 2:
If ("PCA1">30 and "soil moisture content">13) Then "Water"

Rule 3:
If ("PCA1"<-10 and "PCA3">10 and "vegetation coverage"<0.4 and "texture">106)
Then "Habitation"

Rule 4:
If ("vegetation coverage">0.4 and "texture">95 and "PCA1"<-20) Then "Cultivated land"

Rule 5:
If ("PCA3"<-5 and 5<"soil moisture content"<8 and "texture"<50 and "land surface temperature">16)
Then "Bared carbonate rock"

FIGURE 9.3 The final rules after verification and post-processing.

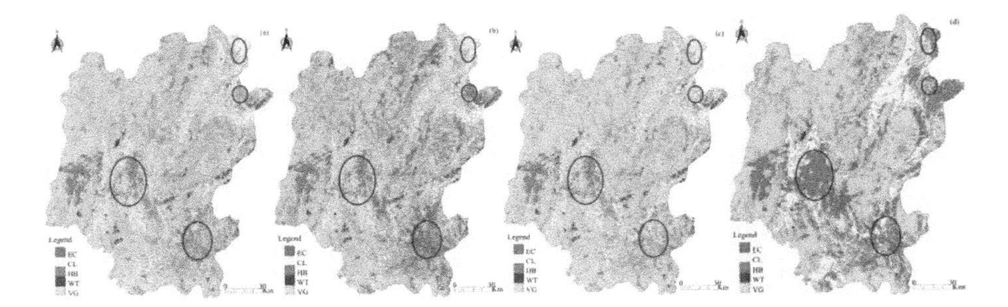

FIGURE 9.4 (a) The results of classification in first test area by the proposed method (Guilin, China); (b) the results of classification in first test area by CL-DT (Guilin, China); (c) the results of classification in first test area by SAE; (d) the results of classification in first test area by traditional DT (CART).

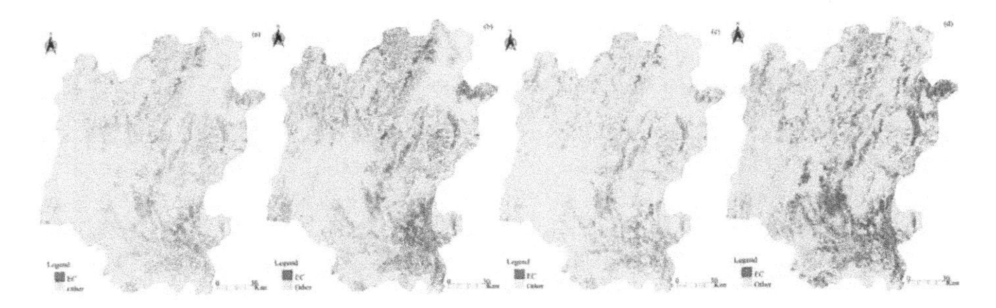

FIGURE 9.5 (a) The extraction results of exposed carbonate by the proposed method (Guilin, China); (b) the extraction results of exposed carbonate by CL-DT (Guilin, China); (c) the extraction results of exposed carbonate by SAE; (d) the extraction results of exposed carbonate by traditional DT (CART).

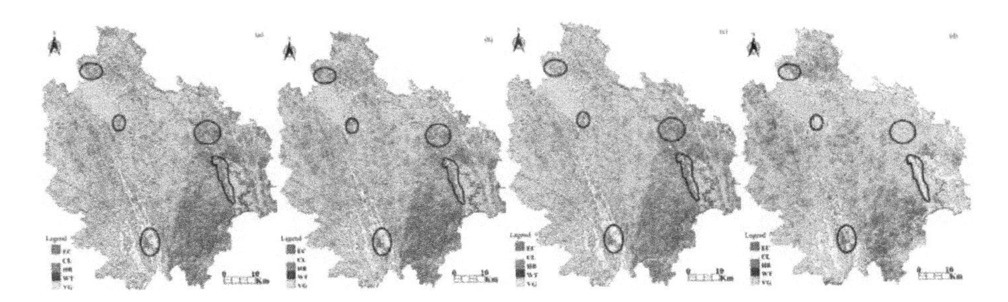

FIGURE 9.6 (a) The results of classification in second test area by the proposed method; (b) the results of classification in second test area by CL-DT; (c) the results of classification in second test area by SAE; (d) the results of classification in second test area by traditional DT (CART).

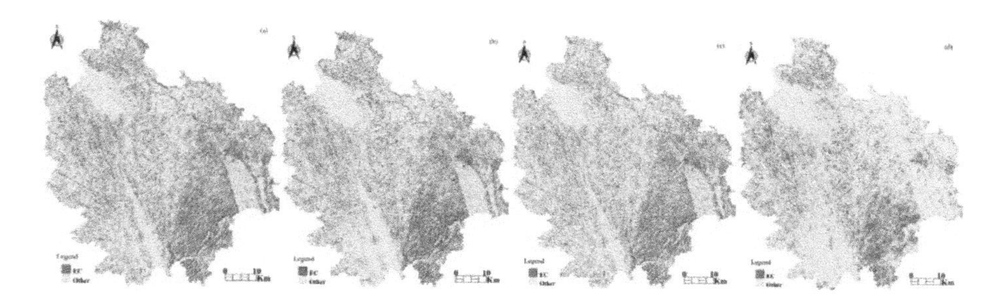

FIGURE 9.7 (a) The extraction results of exposed carbonate by the proposed method (Du'an, China); (b) the extraction results of exposed carbonate by CL-DT (Du'an, China); (c) the extraction results of exposed carbonate by SAE; (d) the extraction results of exposed carbonate by traditional DT (CART).

9.4 COMPARISON ANALYSIS AND VALIDATION IN THE FIELD

9.4.1 Classification Accuracy Comparison

In order to analyze the classification accuracy of the proposed MVU-based CL-DT method, we take classification results using SAE method as a baseline, that is, "true value," and perform the change detection statistical analysis by using ENVI 4.8 software between classification results retrieved by SAE and MVU-based CL-DT, between classification results retrieved by SAE and CL-DT, and between classification results retrieved by SAE and traditional DT, respectively. Compared to the baseline classification results, the relative difference (RD) of classification results and relative accuracy (RA) for two test areas are shown in Table 9.7 and Table 9.8, respectively.

To further analyze the classification accuracy, the proportions of five categories are statistically analyzed and shown in Figure 9.8, As observed to Figure 9.8, the proportion of each category classified by SAE and MVU-based CL-DT are very close. For example, in first test area, the smallest difference between them is about 0.2% for WT, and the biggest difference between them is 1.1% for EC.

As observed in the Table 9.7, Table 9.8, Figure 9.8, and Figure 9.9, the proposed MVU-based CL-DT has the highest classification accuracy relative to baseline classification results. CL-DT takes second place, and traditional DT is the worst.

To further explain the impact of spatial relationship for classification accuracy, spectral curves and magnified windows of classification results, which are retrieved by the MVU-based CL-DT, CL-DT, and traditional DT, are employed for a visual check (see Figure 9.10 and Figure 9.11). As shown in Figure 9.10, although there are three different features, their spectral curves are very close and similar. One of them, pine (masson pine) is the main natural vegetation in first test area, which is recoded as "VG" in classification results. The grass family (for example, rice) and walnuts are the main crop, which is recoded as "CL" in classification results. Because their spectral characteristics are so close and similar, DT misclassifies "VG" and "CL."

TABLE 9.7
The RD of Classification Results and RA in First Test Area

#	SAE (Baseline)	DT	CL-DT	MVU-based CL-DT
RD (%)	0	25.32	16.38	8.93
RA (%)	100	74.68	83.62	91.07

TABLE 9.8
The RD of Classification Results and RA in Second Test Area

#	SAE (Baseline)	DT	CL-DT	MVU-based CL-DT
RD (%)	0	25.03	14.72	8.46
RA (%)	100	74.97	85.28	91.54

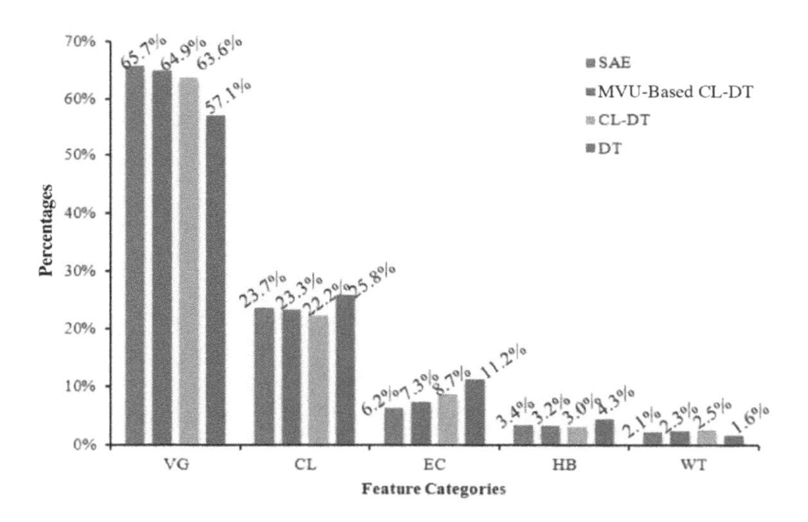

FIGURE 9.8 Comparison of the proportion of categories retrieved by different methods in the first test area.

For example, as observed in Figure 9.11, the classification results of SAE and MVU-based CL-DT are the closest to the real image, and the classification results of traditional DT are significantly different from real image.

As mention in section 9.3.1, the motivation of the CL-DT is to decrease misclassification, which is caused by images with different objects but the same spectra, by using spatial relationship, that is, co-location relationship between instances. Thus,

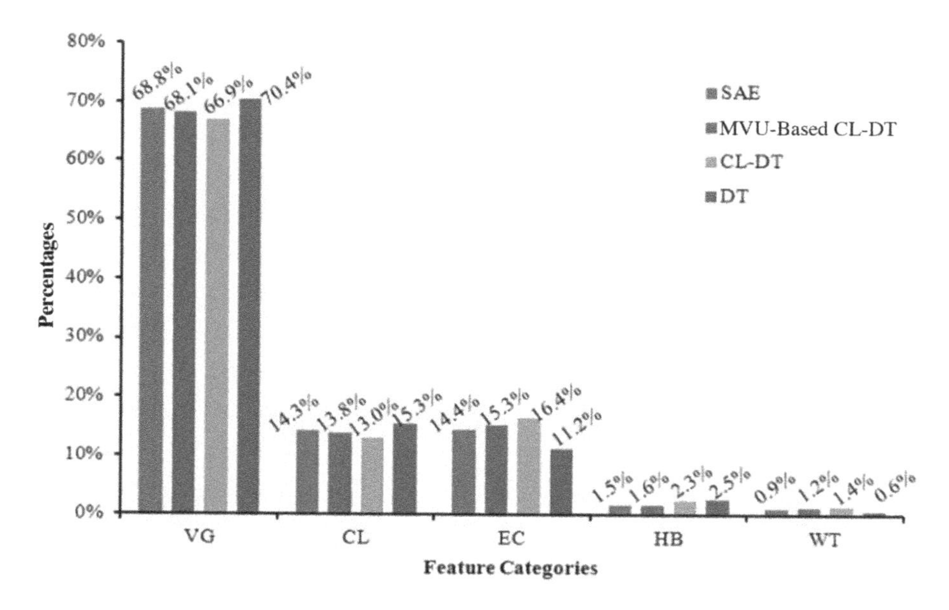

FIGURE 9.9 Comparison of the proportion of categories retrieved by different methods in the second test area.

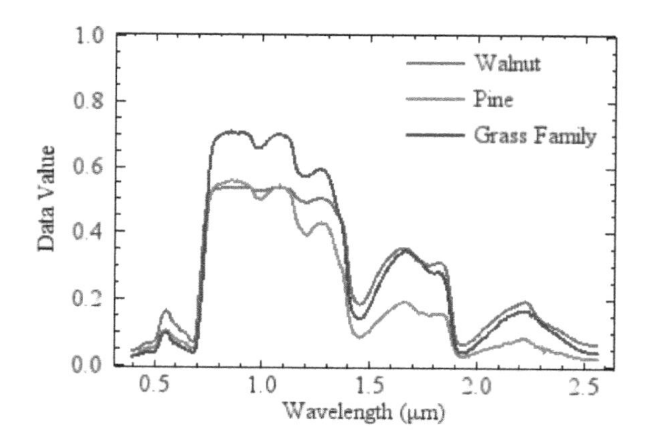

FIGURE 9.10 Spectral curves of features. (Note: These spectral curves of features are from spectral libraries of ENVI 4.8).

MVU-based CL-DT and CL-DT have higher classification accuracy than traditional DT.

On the other hand, through using ground truth ROIs acquired from Google Earth and the confusion matrix method, we can get the produce accuracy (Prod. Acc.), user accuracy (User Acc.), over-accuracy (OA), and Kappa coefficient of four types of method (see Table 9.9, 9.10, and 9.11).

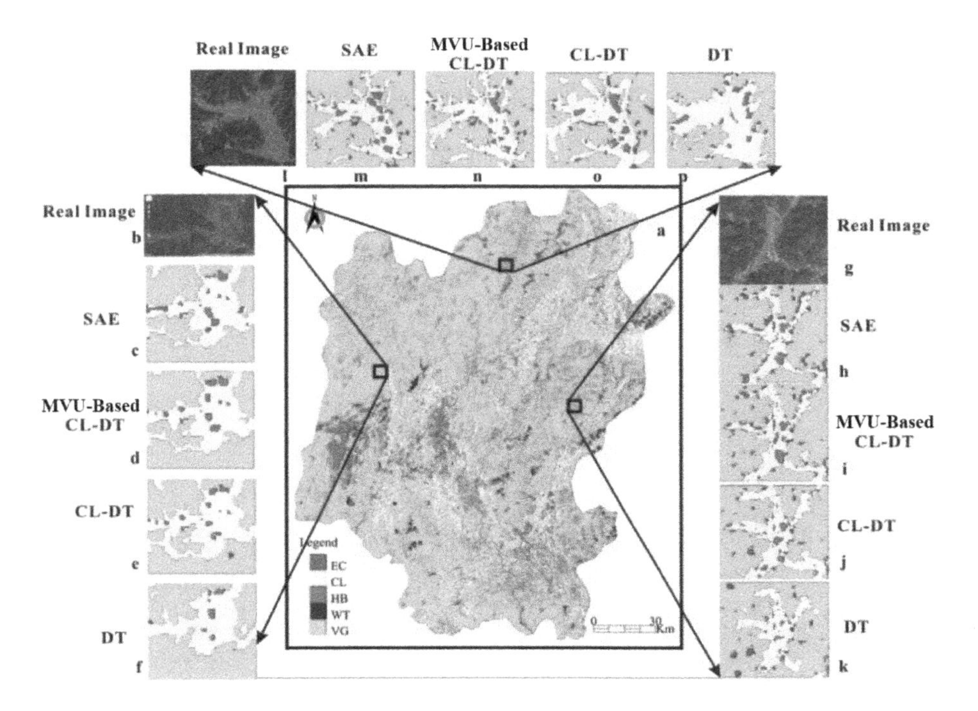

FIGURE 9.11 Magnified windows of classification results in first test area.

(a) The classification results of SAE.
(b)/(g)/(l) True image.
(c)/(h)/(m) Magnified windows of classification results of SAE.
(d)/(i)/(n) Magnified windows of classification results of MVU-based CL-DT.
(e)/(j)/(o) Magnified windows of classification results of CL-DT.
(f)/(k)/(p) Magnified windows of classification results of traditional DT (CART).

TABLE 9.9
Prod. Acc., User Acc., and AA for Different Methods in First Area

#	Prod. Acc.				User Acc.			
	CRAT	CL-DT	SAE	Our method	CRAT	CL-DT	SAE	Our method
VG	100.0%	86.36%	99.55%	96.74%	90.91%	86.36%	100.0%	86.41%
WT	72.82%	89.32%	100.0%	94.50%	98.68%	92.00%	99.69%	100.0%
EC	79.31%	79.31%	80.88%	83.91%	57.98%	71.13%	93.61%	84.88%
HB	84.71%	75.29%	94.65%	95.83%	76.60%	92.75%	96.63%	100.0%
CL	63.39%	83.93%	97.63%	86.07%	81.61%	77.69%	100%	86.78%
AA	80.05%	82.84%	94.54%	91.41%	81.16%	83.99%	97.99%	91.61%

TABLE 9.10
Prod. Acc., User Acc., and AA for Different Methods in Second Area

#	Prod. Acc.				User Acc.			
	CRAT	**CL-DT**	**SAE**	**Our method**	**CRAT**	**CL-DT**	**SAE**	**Our method**
VG	95.28%	86.40%	100.0%	91.75%	94.87%	74.22%	96.28%	89.00%
WT	95.41%	90.91%	100.0%	96.70%	100.0%	89.29%	100.0%	100.0%
EC	50.00%	75.28%	95.37%	80.90%	71.97%	81.71%	83.74%	84.71%
HB	71.33%	81.93%	98.60%	92.94%	100.0%	89.47%	98.60%	95.18%
CL	98.15%	75.00%	85.26%	88.29%	52.48%	78.36%	97.59%	83.76%
AA	82.03%	81.90%	95.85%	90.12%	83.86%	82.61%	95.24%	90.53%

TABLE 9.11
Comparison of AA and Kappa Coefficient in Two Test Areas

#	1st Test Area				2nd Test Area			
	CRAT	**CL-DT**	**SAE**	**Our method**	**CRAT**	**CL-DT**	**SAE**	**Our method**
OA	79.87%	83.3%	98.0%	91.1%	80.33%	81.8%	95.53%	90.1%
Kappa Coefficient	0.75	0.79	0.97	0.89	0.75	0.77	0.94	0.87

9.4.2 Parameters and Computation Time Comparison

To further evaluate the quality of the proposed method, the induced DT parameters and computational time are compared among the MVU-based CL-DT, CL-DT, and DT.

9.4.2.1 Comparison of the Induced Decision Tree Parameters

This chapter also compares the induced DT parameters for the proposed method, CL-DT method, and traditional DT. The compared results are listed in Table 9.12. The total number of nodes, number of leaf nodes, and number of levels of the proposed method decrease by 48%, 45%, and 25%, respectively, compared to CL-DT. The total number of nodes, number of leaf nodes, and number of levels of the proposed method decrease by 56%, 54%, and 33%, respectively, compared to traditional DT. This means that the proposed MVU-based CL-DT algorithm can make a better DT.

9.4.2.2 Comparison of the Computational Time

Another comparison is the running time. The running time of the proposed method consists of two parts, that is, the time of preprocessing (that is, unfolding data sets) and the time of constructing DT. In the preprocessing phase, several minutes are taken to unfold the original data set, that is, large images. However, in the phase of constructing DT, the time can be largely decreased, which has been reported in Table 9.13. As observed in Table 9.13, the necessary time for all of them is largely

TABLE 9.12
Comparison of the Induced Decision Tree Parameters

#	DT	CL-DT	Our method
Total number of nodes	25	21	11
Number of leaf nodes	13	11	6
Number of levels	9	8	6

TABLE 9.13
Comparison of the Computational Time of Constructing DT

#	Time taken (seconds)		
	DT	CL-DT	Our method
Data processing	6	4	3
Decision tree growing	11	8	5
Decision tree drawing	22	18	10
Generating rules	52	37	21

FIGURE 9.12 Validation in field.

decreased. The time taken for rule generation decreases by 43% and 59%, respectively, compared to CL-DT and traditional DT. This demonstrates that the proposed method has higher computation speed in constructing decision tree.

9.4.3 Validation in Field

For the first test area, we conducted a field validation in Guanyang County, Guilin, China (Figure 9.12). We used a Magellan 210 GPS to collect the longitudes

TABLE 9.14
Sample Points

#	Latitude	Longitude	#	Latitude	Longitude
1	25°36′11.52″	111°6′38.88″	11	25°36′39.45″	111°6′49.18″
2	25°36′20.29″	111°6′47.49″	12	25°36′38.63″	111°6′53.45″
3	25°36′17.86″	111°6′39.42″	13	25°36′32.58″	111°6′47.21″
4	25°36′25.42″	111°6′38.86″	14	25°36′39.41″	111°6′57.76″
5	25°36′26.88″	111°6′41.03″	15	25°36′33.75″	111°6′44.85″
6	25°36′25.66″	111°6′1.01″	16	25°36′37.07″	111°6′51.73″
7	25°36′33.95″	111°6′48.99″	17	25°36′40.58″	111°6′32.31″
8	25°36′35.41″	111°6′51.41″	18	25°36′41.16″	111°6′36.61″
9	25°36′37.85″	111°6′48.72″	19	25°36′32.56″	111°6′37.25″
10	25°36′36.39″	111°6′47.11″	20	25°36′25.92″	111°6′45.42″

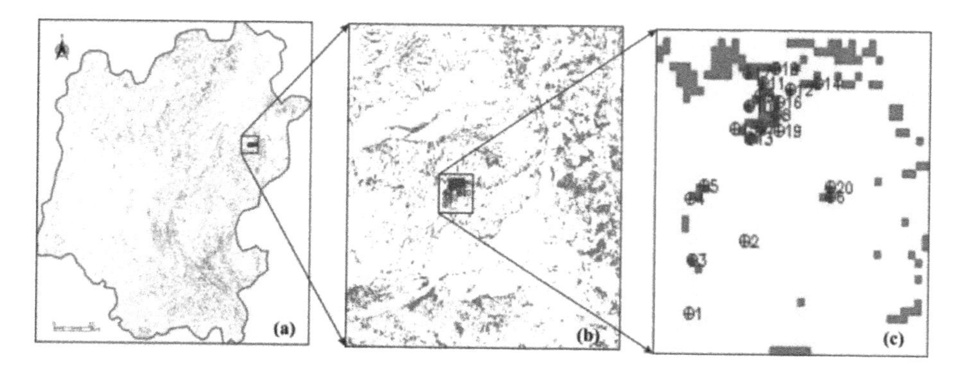

FIGURE 9.13 Validation in the field using points collected by a Magellan 210 GPS.

and latitudes of 20 points that represent the locations of exposed carbonate (Table 9.14).

These collected points are used to validate the extraction results of the exposed carbonate obtained by the proposed method (Figure 9.13). Only two collected points, points 1 and 2, do not match the extraction results, for an accuracy of 90%.

9.5 CONCLUSIONS

The primary contribution of this research is to propose a MVU-based CL-DT algorithm. The algorithm overcomes the deficiency of the traditional CL-DT method, where the Euclidean distance of instances that are nonlinear distributions in high dimensional space cannot accurately reflect the co-location relationship between instances through merging the maximum variance unfolding algorithm with the

CL-DT. In this way, the issue of "the different objects but the same/similar spectrum" in remote sensing images, which decreases the classification accuracy, can be better solved.

This chapter has provided detailed descriptions of algorithms and steps. First, the MVU algorithm is utilized to unfold the input data, and the unfolded distances between instances in unfolded data are calculated. Second, according to the unfolded distances, the *RRS* between instances is determined. Third, MVU-based co-location instances are found on the basis of the *RRS*. Then the distinct events are determined and MVU-based co-location rules are generated. Finally, the MVU-based co-location rules are merged into the DT to generate the MVU-based co-location decision rules.

The proposed method has been used to classify the remote sensing images in two test areas that are typical karst rocky desertification areas. Compared to SAE, which is the baseline, CL-DT, and traditional DT (CART), it demonstrates that (1) the proposed method has the highest classification accuracy, with relative accuracy of 91.07% and 91.54% in two test areas, respectively, relative to baseline. However, the CL-DT method reaches 83.62% and 85.28%, respectively, and traditional DT only gets 74.68% and 74.97%, respectively; and (2) the proposed method can produce a better tree, because the total number of nodes, the number of leaf nodes, and the number of levels of the proposed method decreases by 48%, 45%, and 25%, respectively, compared to CL-DT, decreases by 56%, 54%, 33%, respectively, compared to traditional DT; and the time taken for data processing, decision tree generation, drawing the tree, and generating rules decreases by 25%, 38%, 44%, and 43%, respectively, compared to CL-DT, and decreases by 50%, 55%, 55%, 60%, respectively, compared to traditional DT. With the calculation results of the confusion matrix using the ROIs, it can be concluded that (1) the OAs for MVU-based CL-DT reach 91.10% and 90.53% and Kappa coefficients are 0.89 and 0.87, in two areas, respectively; (2) the OAs for SAE achieve 98.10% and 95.53%, and Kappa coefficients are 0.97 and 0.94 in two areas, respectively; (3) the OAs for CL-DT achieve 83.30% and 81.80%, and Kappa coefficients are 0.79 and 0.77 in two areas, respectively; (4) the OAs for DT reach 79.87% and 80.33%, and Kappa coefficients are 0.75 and 0.75 in two areas, respectively.

REFERENCES

Breiman, L., Friedman, J.H., Olshen, R.A., and Stone, C.J., Classification and regression trees. Belmont, CA, USA: Wadsworth, 1984.

Chen, Y.S., Lin, Z.H., Zhao, X., Wang, G., and Gu, Y.F., Deep learning-based classification of hyperspectral data. *IEEE Journal of Selected Topics in Applied Earth Observations and Remote Sensing*, vol. 7, no. 6, 2014, pp. 2094–2107.

Greiner, G., and Hormann, K., Efficient clipping of arbitrary polygons. *ACM Transactions on Graphics*, vol. 17, no. 2, April 1998, pp. 71–83.

Kanazawaa, Y., and Kanatani, K., Image mosaicing by stratified matching. *Image and Vision Computing*, vol. 22, 2004, pp. 93–103.

Kardoulas, N.G., Bird, A.C., and Lawan, A.I., Geometric correction of spot and Landsat imagery: A comparison of map-and GPS-derived control points. *American Society for Photogrammetry and Remote Sensing*, vol. 62, no. 10, 1996, pp. 1171–1177.

Liu, P.J., Zhang, L., and Kurban, A., A method for monitoring soil water contents using satellite remote sensing. *Journal of Remote Sensing*, vol. 1, no. 2, 1997, pp. 135–139.

Mohammad, A., Shi, Z., and Ahmad, Y., Application of GIS and remote sensing in soil degradation assessments in the Syrian coast. *Journal of Zhejiang University (Agric. & Life Sci.)*, vol. 26, no. 2, 2002, pp. 191–196.

Qin, Z., and Karniell, A., A mono-window algorithm for retrieving land surface temperature from Landsat TM data and its application to the Israel–Egypt border region. *International Journal of Remote Sensing*, vol. 22, no. 18, 2001, pp. 3719–3746.

Storey, J.C., and Choate, M.J., Landsat-5 Bumper-mode geometric correction. *IEEE Transactions on Geoscience and Remote Sensing*, vol. 42, no. 12, 2004, pp. 2695–2703.

Vincent, P., Larochelle, H., Lajoie, I., Bengio, Y., and Manzagol, P., Stacked denoising autoencoders: Learning useful representations in a deep network with a local Denoising criterion. *The Journal of Machine Learning Research*, vol. 11, 2010, pp. 3371–3408.

Yang, W.J., The registration and mosaic of digital image remotely sensed. *Proceedings of the 11th Asian Conference on Remote Sensing*, vol. 2, 1990, pp. ACRSQ–12–1–6.

Zhou, G., Co-location decision tree for enhancing decision-making of pavement maintenance and rehabilitation. *Ph.D. dissertation*, Virginia Tech, Blacksburg, Virginia, USA, 2011.

Zhou, G., and Wang, L., Co-location decision tree for enhancing decision-making of pavement maintenance and rehabilitation. *Transportation Research Part C: Emerging Technologies*, vol. 21, no. 1, 2012, pp. 287–305.

Zhou, G., Wang, L., Wang, D., and Reichle, S., Integration of GIS and data mining technology to enhance the pavement management decision making. *Journal of Transportation Engineering*, vol. 136, no. 4, 2010, pp. 332–341.

Index

Note: Page numbers in *italics* refer to figures; those in **bold** refer to tables.

O

P

R

S